The Ascent of Man

The Ascent of Man

Jacob Bronowski

This book is published to accompany the television series entitled *The Ascent of Man*, first broadcast on BBC in 1973.

Series Editor: Adrian Malone
Producer: Richard Gilling

11

This paperback edition published in 2011 by BBC Books, an imprint of Ebury Publishing.
A Random House Group Company

First published in hardback by the British Broadcasting Corporation in 1973
First paperback edition published in 1976

The Random House Group Limited Reg. No. 954009

Addresses for companies within the Random House Group can be found at www.randomhouse.co.uk

A CIP catalogue record for this book is available from the British Library.

ISBN 978 1849901154

Project editor: Nicholas Payne
Designer: O'Leary & Cooper
Production: David Brimble

Printed and bound in Great Britain by Clays Ltd, St Ives plc

THE ASCENT OF MAN

Dr Bronowski's magnificent thirteen-part BBC television series *The Ascent of Man* traced our rise – both as a species and as moulders of our own environment and future. The book of the programmes covers the history of science, but of science in the broadest terms. Invention from the flint tool to geometry, from the arch to the theory of relativity, are shown to be expressions of man's specific ability to understand nature, to control it, not to be controlled by it. Dr Bronowski's rare grasp not only of science, but also of its historical and social context, gave him great advantages as an historian of ideas. The book gives us a new perspective not just on science, but on civilisation.

Dr Jacob Bronowski, who was born in Poland in 1908, died in 1974. His family had settled in Britain and he was educated at Cambridge University.

He was distinguished not only as a scientist but also as the author of books and broadcasts on the arts. Many viewers will remember his science programmes on television: he also wrote radio plays, including one which won the Italia Prize.

Dr Bronowski, who was an Honorary Fellow of Jesus College, Cambridge, had lived and worked in America since 1964, as a Senior Fellow and Director of the Council for Biology in Human Affairs at the Salk Institute for Biological Studies, San Diego, California.

Other Books by J. Bronowski

The Poet's Defence 1939 & 1966
William Blake and The Age of Revolution 1944 & 1965
The Common Sense of Science 1951
The Face of Violence 1954 & 1967
Science and Human Values 1958
with *The Abacus and The Rose:*
A New Dialogue on Two World Systems 1965
Selections from William Blake 1958
The Western Intellectual Tradition
(with Prof Bruce Mazlish) 1960
Insight 1964
The Identity of Man 1965 & 1972
Nature and Knowledge:
The Philosophy of Contemporary Science 1969

CONTENTS

FOREWORD
by Richard Dawkins

'Last renaissance man' has become a cliché, but we forgive a cliché on the rare occasion when it is true. Certainly it is hard to think of a better candidate for the accolade than Jacob Bronowski. You'll find other scientists who can parade a deep parallel knowledge of the arts, or – in one actual case – combine eminence in science with pre-eminence in Chinese history. But who more than Bronowski weaves a deep knowledge of history, art, cultural anthropology, literature and philosophy into one seamless cloth with his science? And does it lightly, effortlessly, never sinking to pretension? Bronowski uses the English language – not his first language, which makes it all the more remarkable – as a painter uses his brush, with mastery all the way from broad canvas to exquisite miniature.

Inspired by the *Mona Lisa*, here is what he has to say about arguably the first and greatest renaissance man, whose drawing of the baby in the womb introduced the television version of *The Ascent of Man*:

> Man is unique not because he does science, and he is unique not because he does art, but because science and art equally are expressions of his marvellous plasticity of mind. And the *Mona Lisa* is a very good example, because after all what did Leonardo

> do for much of his life? He drew anatomical pictures, such as the baby in the womb in the Royal Collection at Windsor. And the brain and the baby is exactly where the plasticity of human behaviour begins.

How deftly Bronowski segues from Leonardo's drawing to the Taung baby: type-specimen of our ancestral genus *Australopithecus*, victim – as we now know, though Bronowski didn't when he performed his mathematical analysis on the tiny skull – of a giant eagle two million years ago.

There's a quotable aphorism on every page of this book, something to treasure, something to stick on your door for all to see, an epitaph, perhaps, for the gravestone of a great scientist. 'Knowledge ... is an unending adventure at the edge of uncertainty.' Uplifting? Yes. Inspiring? Without doubt. But read it in context and it is shocking. The grave turns out to belong to an entire tradition of European scholarship, destroyed by Hitler and his allies almost overnight:

> Europe was no longer hospitable to the imagination – and not just the scientific imagination. A whole conception of culture was in retreat: the conception that human knowledge is personal and responsible, an unending adventure at the edge of uncertainty. Silence fell, as after the trial of Galileo. The great men went out into a threatened world. Max Born. Erwin Schrödinger. Albert Einstein. Sigmund Freud. Thomas Mann. Bertolt Brecht. Arturo Toscanini. Bruno Walter. Marc Chagall.

Words so powerful don't need a raised voice or ostentatious tears. Bronowski's words gained impact from his calm, humane, understated tones, with the engagingly rolled Rs as he looked straight into the camera, spectacles flashing like beacons in the dark.

That was a rare dark passage in a book that is mostly filled with light, and genuinely uplifting. You can hear Bronowski's distinctive voice through this book, and you can see his expressive hand chopping down to cut through complexity and make a point. He stands before a great sculpture, Henry Moore's *The Knife Edge*, to tell us,

> The hand is the cutting edge of the mind. Civilisation is not a collection of finished artefacts, it is the elaboration of processes. In the end, the march of man is the refinement of the hand in action. The most powerful drive in the ascent of man is his pleasure in his own skill. He loves to do what he does well and, having done it well, he loves to do it better. You see it in his science. You see it in the magnificence with which he carves and builds, the loving care, the gaiety, the effrontery. The monuments are supposed to commemorate kings and religions, heroes, dogmas, but in the end the man they commemorate is the builder.

Bronowski was a rationalist and an iconoclast. He was not content to bask in the achievements of science but sought to provoke, to pique, to needle.

> That is the essence of science: ask an impertinent question, and you are on the way to a pertinent answer.

That applies not just to science but to all learning, epitomised, for Bronowski by one of the world's oldest and greatest universities – in Germany as it happens:

> The University is a Mecca to which students come with something less than perfect faith. It is important that students

> bring a certain ragamuffin, barefoot irreverence to their studies; they are not here to worship what is known but to question it.

Bronowski treated the magical speculations of primitive man with sympathy and understanding, but in the end

> … magic is only a word, not an answer. In itself, magic is a word which explains nothing.

There is magic – the right kind of magic – in science. There is poetry too, and magical poetry on every page of this book. Science is the poetry of reality. If he didn't say that, it is the kind of thing he might have said, articulate polymath and gentle sage, whose wisdom and intelligence symbolises all that is best in the ascent of man.

INTRODUCTION

The first outline of *The Ascent of Man* was written in July 1969 and the last foot of film was shot in December 1972. An undertaking as large as this, though wonderfully exhilarating, is not entered lightly. It demands an unflagging intellectual and physical vigour, a total immersion, which I had to be sure that I could sustain with pleasure; for instance, I had to put off researches that I had already begun; and I ought to explain what moved me to do so.

There has been a deep change in the temper of science in the last twenty years: the focus of attention has shifted from the physical to the life sciences. As a result, science is drawn more and more to the study of individuality. But the interested spectator is hardly aware yet how far-reaching the effect is in changing the image of man that science moulds. As a mathematician trained in physics, I too would have been unaware, had not a series of lucky chances taken me into the life sciences in middle age. I owe a debt for the good fortune that carried me into two seminal fields of science in one lifetime; and though I do not know to whom the debt is due, I conceived *The Ascent of Man* in gratitude to repay it.

The invitation to me from the British Broadcasting Corporation was to present the development of science in a series of television programmes to match those of Lord Clark on *Civilisation*. Television is an admirable medium for exposition in several ways: powerful and immediate to the eye, able to take the spectator bodily into the places and processes that are described, and conversational enough to make him conscious that what he witnesses are not events but the actions of people. The last of these merits is to my mind the most cogent, and it weighed most with me in agreeing to cast a personal biography of ideas in the form of television essays. The point is that knowledge in general and science in particular does not consist of abstract but of manmade ideas, all the way from its beginnings to its modern and idiosyncratic models. Therefore the underlying concepts that unlock nature must be shown to arise early and in the simplest cultures of man from his basic and specific faculties. And the development of science which joins them in more and more complex conjunctions must be seen to be equally human: discoveries are made by men, not merely by minds, so that they are alive and charged with individuality. If television is not used to make these thoughts concrete, it is wasted.

The unravelling of ideas is, in any case, an intimate and personal endeavour, and here we come to the common ground between television and the printed book. Unlike a lecture or a cinema show, television is not directed to crowds. It is addressed to two or three people in a room, as a conversation face to face – a one-sided conversation for the most part, as the book is, but homely and Socratic nevertheless. To me, absorbed in the philosophic undercurrents of knowledge, this is the most attractive gift of television, by which it may yet become as persuasive an intellectual force as the book.

The printed book has one added freedom beyond this: it is not remorselessly bound to the forward direction of time, as any spoken discourse is. The reader can do what the viewer and the listener cannot, which is to pause and reflect, turn the pages back and the argument over, compare one fact with another and, in general, appreciate the detail of evidence without being distracted by it. I have taken advantage of this more leisurely march of mind whenever I could, in putting on paper now what was first said on the television screen. What was said had required a great volume of research, which turned up many unexpected links and oddities, and it would have been sad not to capture some of that richness in this book. Indeed, I should have liked to do more, and to interleave the text in detail with the source material and quotations on which it rests. But that would have turned the book into a work for students instead of the general reader.

In rendering the text used on the screen, I have followed the spoken word closely, for two reasons. First, I wanted to preserve the spontaneity of thought in speech, which I had done all I could to foster wherever I went. (For the same reason, I had chosen whenever possible to go to places that were as fresh to me as to the viewer.) Second and more important, I wanted equally to guard the spontaneity of the argument. A spoken argument is informal and heuristic; it singles out the heart of the matter and shows in what way it is crucial and new; and it gives the direction and line of the solution so that, simplified as it is, still the logic is right. For me, this philosophic form of argument is the foundation of science, and nothing should be allowed to obscure it.

The content of these essays is in fact wider than the field of science, and I should not have called them *The Ascent of Man* had I not had in mind other steps in our cultural evolution too. My ambition here has been the same as in my other books, whether in literature or in science: to create a philosophy for the twentieth century which shall be all of one piece. Like them, this series presents a philosophy rather than a history, and a philosophy of nature rather than of science. Its subject is a contemporary version of what used to be called Natural Philosophy. In my view, we are in a better frame of mind today to conceive a natural philosophy than at any time in the last three hundred years. This is because the recent findings in human biology have given a new direction to scientific thought, a shift from the general to the individual, for the first time since the Renaissance opened the door into the natural world.

There cannot be a philosophy, there cannot even be a decent science, without humanity. I hope that sense of affirmation is manifest in this book. For me, the understanding of nature has as its goal the understanding of human nature, and of the human condition within nature.

To present a view of nature on the scale of this series is as much an experiment as an adventure, and I am grateful to those who made both possible. My first debt is to the Salk Institute for Biological Studies which has long supported my work on the subject of human specificity, and which gave me a year of sabbatical leave to film the programmes. I am greatly indebted also to the British Broadcasting Corporation and its associates, and very particularly there to Aubrey Singer who invented the massive theme and urged it on me for two years before I was persuaded.

The list of those who helped to make the programmes is so long that I must put it on a page of its own, and thank them in a body; it was a pleasure to work with them. However, I cannot pass over the names of the producers that stand at the head of the list, and particularly Adrian Malone and Dick Gilling, whose imaginative ideas transubstantiated the word into flesh and blood.

Two people worked with me on this book, Josephine Gladstone and Sylvia Fitzgerald, and did much more; I am happy to be able to thank them here for their long task. Josephine Gladstone had charge of all the research for the series since 1969, and Sylvia Fitzgerald helped me plan and prepare the script at each successive stage. I could not have had more stimulating colleagues.

J. B.
La Jolla, California
August 1973

CHAPTER ONE

LOWER THAN THE ANGELS

Man is a singular creature. He has a set of gifts which make him unique among the animals: so that, unlike them, he is not a figure in the landscape – he is a shaper of the landscape. In body and in mind he is the explorer of nature, the ubiquitous animal, who did not find but has made his home in every continent.

It is reported that when the Spaniards arrived overland at the Pacific Ocean in 1769 the California Indians used to say that at full moon the fish came and danced on these beaches. And it is true that there is a local variety of fish, the grunion, that comes up out of the water and lays its eggs above the normal high-tide mark. The females bury themselves tail first in the sand and the males gyrate round them and fertilise the eggs as they are being laid. The full moon is important, because it gives the time needed for the eggs to incubate undisturbed in the sand, nine or ten days, between these very high tides and the next ones that will wash the hatched fish out to sea again.

Every landscape in the world is full of these exact and beautiful adaptations, by which an animal fits into its environment like one cog-wheel into another. The sleeping hedgehog waits for the spring to burst its metabolism into life. The humming-bird beats the air and dips its needle-fine beak into hanging blossoms. Butterflies mimic leaves and even noxious creatures to deceive their predators. The mole plods through the ground as if he had been designed as a mechanical shuttle.

So millions of years of evolution have shaped the grunion to fit and sit exactly with the tides. But nature – that is, biological evolution – has not fitted man to any specific environment. On the contrary, by comparison with the grunion he has a rather crude survival kit; and yet – this is the paradox of the human condition – one that fits him to all environments. Among the multitude of animals which scamper, fly, burrow and swim around us, man is the only one who is not locked into his environment. His imagination, his reason, his emotional subtlety and toughness, make it possible for him not to accept the environment but to change it. And that series of inventions, by which man from age to age has remade his environment, is a different kind of evolution – not biological, but cultural evolution. I call that brilliant sequence of cultural peaks *The Ascent of Man.*

I use the word ascent with a precise meaning. Man is distinguished from other animals by his imaginative gifts. He makes plans, inventions, new discoveries, by putting different talents together; and his discoveries become more subtle and penetrating, as he learns to combine his talents in more complex and intimate ways. So the great discoveries of different ages and different cultures, in technique, in science, in the arts, express in their progression a richer and more intricate conjunction of human faculties, an ascending trellis of his gifts.

Of course, it is tempting – very tempting to a scientist – to hope that the most original achievements of the mind are also the most recent. And we do indeed have cause to be proud of some modern work. Think of the unravelling of the code of heredity in the DNA spiral; or the work going forward on the special faculties of the human brain. Think of the philosophic insight that saw into the Theory of Relativity or the minute behaviour of matter on the atomic scale.

Yet to admire only our own successes, as if they had no past (and were sure of the future), would make a caricature of knowledge. For human achievement, and science in particular, is not a museum of finished constructions. It is a progress, in which the first experiments of the alchemists also have a formative place, and the sophisticated arithmetic that the Mayan astronomers of Central America invented for themselves independently of the Old World. The stonework of Machu Picchu in the Andes and the geometry of the Alhambra in Moorish Spain seem to us, five centuries later, exquisite works of decorative art. But if we stop our appreciation there, we miss the originality of the two cultures that made them. Within their time, they are constructions as arresting and important for their peoples as the architecture of DNA for us.

In every age there is a turning-point, a new way of seeing and asserting the coherence of the world. It is frozen in the statues of Easter Island that put a stop to time – and in the medieval clocks in Europe that once also seemed to say the last word about the heavens for ever. Each culture tries to fix its visionary moment, when it was transformed by a new conception either of nature or of man. But in retrospect, what commands our attention as much are the continuities – the thoughts that run or recur from one civilisation to another. There is nothing in modern chemistry more unexpected than putting together alloys with new properties; that was discovered after the time of the birth of Christ in South America, and long before that in Asia. Splitting and fusing the atom both derive, conceptually, from a discovery made in prehistory: that stone and all matter has a structure along which it can be split and put together in new arrangements. And man made biological inventions almost as early: agriculture – the domestication of wild wheat, for example – and the improbable idea of taming and then riding the horse.

In following the turning-points and the continuities of culture, I shall follow a general but not a strict chronological order, because what interests me is the history of man's mind as an unfolding of his different talents. I shall be relating his ideas, and particularly his scientific ideas, to their origins in the gifts with which nature has endowed man, and which make him unique. What I present, what has fascinated me for many years, is the way in which man's ideas express what is essentially human in his nature.

So these programmes or essays are a journey through intellectual history, a personal journey to the high points of man's achievement. Man ascends by discovering the fullness of his own gifts (his talents or faculties) and what he creates on the way are monuments to the stages in his understanding of nature and of self – what the poet W. B. Yeats called 'monuments of unageing intellect'.

Where should one begin? With the Creation – with the creation of man himself. Charles Darwin pointed the way with *The Origin of Species* in 1859, and then in his book of 1871, *The Descent of Man*. It is almost certain now that man first evolved in Africa near the equator. Typical of the places where his evolution may have begun is the savannah country that stretches out across Northern Kenya and South West Ethiopia near Lake Rudolf. The lake lies in a long ribbon north and south along the Great Rift Valley, hemmed in by over four million years of thick sediments that settled in the basin of what was formerly a much more extensive lake. Much of its water comes by way of the winding, sluggish Omo. For the origins of man, this is a possible area: the valley of the river Omo in Ethiopia near Lake Rudolf.

The ancient stories used to put the creation of man into a golden age and a beautiful, legendary landscape. If I were telling the story of Genesis now, I should be standing in the Garden of Eden. But this is manifestly not the Garden of Eden. And yet I am at the navel of the world, at the birthplace of man, here in the East African Rift Valley, near the equator. The slumped levels in the Omo basin, the bluffs, the barren delta, record a historic past of man. And if this ever was a Garden of Eden, why, it withered millions of years ago.

I have chosen this place because it has a unique structure. In this valley was laid down, over the last four million years, layer upon layer of volcanic ash, interbedded with broad bands of shale and mudstone. The deep deposit was formed at different times, one stratum after another, visibly separated according to age: four million years ago, three million years ago, over two million years ago, somewhat under two million years ago. And then the Rift Valley buckled it and stood it on end, so that now it makes a map in time, which we see stretching into the distance and the past. The record of time in the strata, which is usually buried underfoot, has been tip-tilted in the cliffs that flank the Omo, and spread out like the ribs of a fan.

These cliffs are the strata on edge: in the foreground the bottom level, four million years old, and beyond that the next lowest, well over three million years old. The remains of a creature like man appear beyond that, and the remains of the animals that lived at the same time.

The animals are a surprise, because it turns out that they have changed so little. When we find in the sludge of two million years ago the fossils of the creature who was to become man, we are struck by the differences between his skeleton and ours – by the development of the skull, for instance. So, naturally, we expect

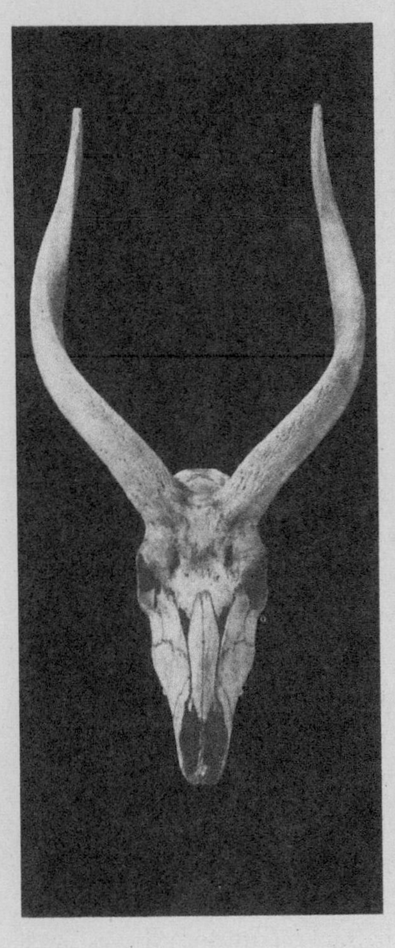

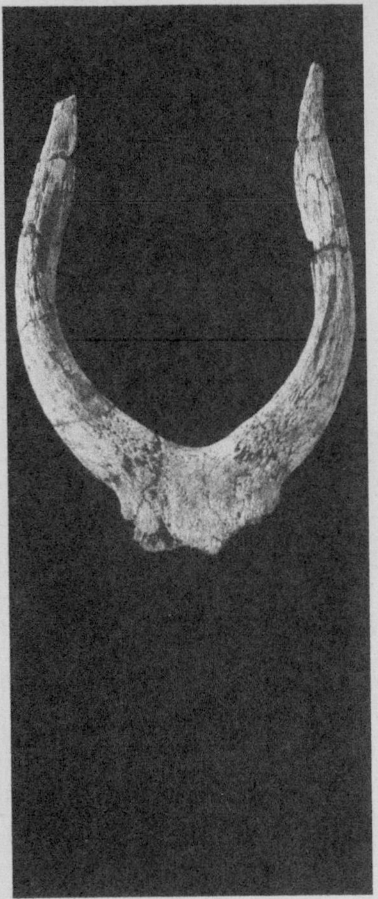

The animals are a surprise, because it turns out that they have changed so little.

Modern and fossil nyala horns from Omo. The fossil horns are over two million years old.

the animals of the savannah also to have changed greatly. But the fossil record in Africa shows that this is not so. Look as the hunter does at the Topi antelope now. The ancestor of man that hunted its ancestor two million years ago would at once recognise the Topi today. But he would not recognise the hunter today, black or white, as his own descendant.

Yet it is not hunting in itself (or any other single pursuit) that has changed man. For we find that among the animals the hunter has changed as little as the hunted. The serval cat is still powerful in pursuit, and the oryx is still swift in flight; both perpetuate the same relation between their species as they did long ago. Human evolution began when the African climate changed to drought: the lakes shrank, the forest thinned out to savannah. And evidently it was fortunate for the forerunner of man that he was not well adapted to these conditions. For the environment exacts a price for the survival of the fittest; it captures them. When animals like Grevy's zebra were adapted to the dry savannah, it became a trap in time as well as space; they stayed where they were, and much as they were. The most gracefully adapted of all these animals is surely Grant's gazelle; yet its lovely leap never took it out of the savannah.

In a parched African landscape like Omo, man first put his foot to the ground. That seems a pedestrian way to begin the Ascent of Man, and yet it is crucial. Two million years ago, the first certain ancestor of man walked with a foot which is almost indistinguishable from the foot of modern man. The fact is that when he put his foot on the ground and walked upright, man made a commitment to a new integration of life and therefore of his limbs.

The one to concentrate on, of course, is the head, because of all human organs it has undergone the most far-reaching and formative changes. Happily, the head leaves a lasting fossil (unlike the soft organs), and though it is less informative about the brain than we should like, at least it gives us some measure of its size. A number of fossil skulls have been found in Southern Africa in the last fifty years which establish the characteristic structure of the head when it began to be man-like. The picture on page 27 shows what it looked like over two million years ago. It is a historic skull, found not at Omo, but south of the equator at a place called Taung, by an anatomist called Raymond Dart. It is a baby, five to six years old, and though the face is nearly complete, part of the skull is sadly missing. In 1924 it was a puzzling find, the first of its kind, and was treated with caution even after Dart's pioneering work on it.

Yet Dart instantly recognised two extraordinary features. One is that the *foramen magnum* (that is, the hole in the skull that the spinal cord comes up through to the brain) is upright; so that this was a child that held its head up. That is one man-like feature; for in the monkeys and apes the head hangs forward from the spine, and does not sit upright on top of it. And the other is the teeth. The teeth are always tell-tale. Here they are small, they are square – these are still the child's milk teeth – they are not the great, fighting canines that the apes have. That means that this was a creature that was going to forage with its hands and not its mouth. The evidence of the teeth also implies that it was probably eating meat, raw meat; and so the hand-using creature was almost certainly making tools, pebble-tools, stone choppers, to carve it and to hunt.

Dart called this creature *Australopithecus*. It is not a name that I like; it just means Southern Ape, but it is a confusing name for an African creature that for the first time was not an ape. I suspect that Dart, who was born in Australia, put a pinch of mischief into his choice of the name.

It took ten years before more skulls were found – adult skulls now – and it was not until late in the 1950s that the story of *Australopithecus*

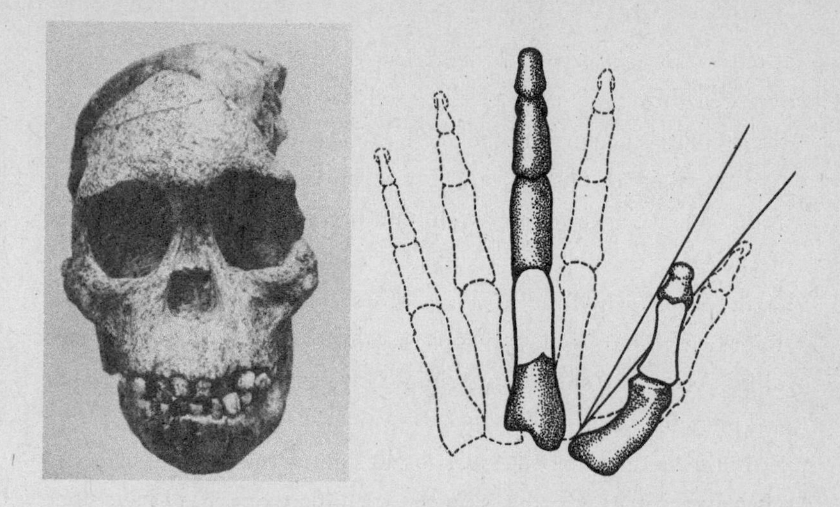

(*Left*) I do not know how the Taung baby began life, but to me it still remains the primordial infant from which the whole adventure of man began.
The Taung child's skull

(*Right*) The ancestor of man had a short thumb, and therefore could not manipulate very delicately.
Finds of finger and thumb bones of Australopithecus *from the lowest beds of Olduvi Gorge superimposed on the bones of a modern hand*

was substantially pieced together. It started in South Africa, then it moved north to Olduvai Gorge in Tanzania, and most recently the richest finds of fossils and tools have turned up in the basin of Lake Rudolf. This history is one of the scientific delights of the century. It is every bit as exciting as the discoveries in physics before 1940, and those in biology since 1950; and it is as rewarding as either of those in the light that it throws on our nature as human beings.

For me, the little *Australopithecus* baby has a personal history. In 1950, when its humanity was by no means accepted, I was asked to do a piece of mathematics. Could I combine a measure of the size of the Taung child's teeth with their shape, so as to discriminate them from the teeth of apes? I had never held a fossil skull in my hands, and I was by no means an expert on teeth. But it worked pretty well; and it transmitted to me a sense of excitement which I remember at this instant. I, at over forty, having spent a lifetime in doing abstract mathematics about the shapes of things, suddenly saw my knowledge reach back two million years and shine a searchlight into the history of man. That was phenomenal.

And from that moment I was totally committed to thinking about what makes man what he is: in the scientific work that I have done since then, the literature that I have written, and in these programmes. How did the hominids come to be the kind of man that I honour: dexterous, observant, thoughtful, passionate, able to manipulate in the mind the symbols of language and mathematics both, the visions of art and geometry and poetry and science? How did the ascent of man take him from those animal beginnings to that rising enquiry into the workings of nature, that rage for knowledge, of which these essays are one expression? I do not know how the Taung baby began life, but to me it still remains the primordial infant from which the whole adventure of man began.

The human baby, the human being, is a mosaic of animal and angel. For example, the reflex that makes the baby kick is already there in the womb – every mother knows that – and it is there in all vertebrates. The reflex is self-sufficient, but it sets the stage for more elaborate movements, which have to be practised before they become automatic. Here by eleven months it urges the baby to crawl. That brings in new movements, and they then lay down and consolidate the pathways in the brain (specifically the cerebellum, where muscular action and balance are integrated) that will form a whole repertoire of subtle, complex movements and make them second nature to him. Now the cerebellum is in control. All that the conscious mind has to do is to issue a command. And by fourteen months the command is 'Stand!' The child has entered the human commitment to walk upright.

Every human action goes back in some part to our animal origins; we should be cold and lonely creatures if we were cut off from that blood-stream of life. Nevertheless, it is right to ask for a distinction: What are the physical gifts that man must share with the animals, and what are the gifts that make him different? Consider any example, the more straightforward the better – say, the simple action of an athlete when running or jumping. When he hears the gun, the starting response of the runner is the same as the flight response of the gazelle. He seems all animal in action. The heartbeat goes up; when he sprints at top speed the heart is pumping five times as much blood as normal, and ninety per cent of it is for the muscles. He needs twenty gallons of air a minute now to aerate his blood with the oxygen that it must carry to the muscles.

The violent coursing of the blood and intake of air can be made visible, for they show up as heat on infra-red films which are sensitive to such radiation. (The blue or light zones are hottest; the red or dark zones are cooler.) The flush that we see and that the infra-red camera analyses is a by-product that signals the limit of muscular action. For the main chemical action is to get energy for the muscles by burning sugar there; but three-quarters of that is lost as heat. And there is another limit, on the runner and the gazelle equally, which is more severe. At this speed, the chemical burn-up in the muscles is too fast to be complete. The waste products of incomplete burning, chiefly lactic acid, now foul up the blood. This is what causes fatigue, and blocks the muscle

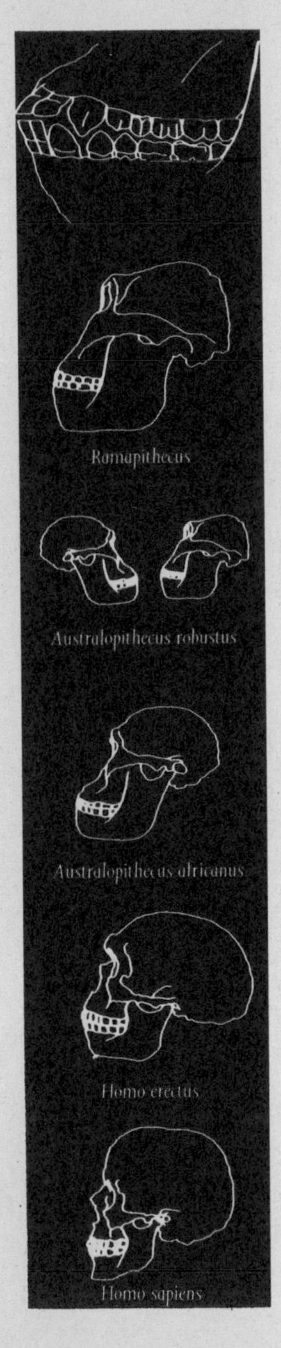

(*Left*) The head is the spring which drives cultural evolution.
Computer-graphic display of stages in evolution of the head

action until the blood can be cleaned with fresh oxygen.

So far, there is nothing to distinguish the athlete from the gazelle – all that, in one way or another, is the normal metabolism of an animal in flight. But there is a cardinal difference: the runner was not in flight. The shot that set him off was the starter's pistol, and what he was experiencing, deliberately, was not fear but exaltation. The runner is like a child at play; his actions are an adventure in freedom, and the only purpose of his breathless chemistry was to explore the limits of his own strength.

Naturally there are physical differences between man and the other animals, even between man and the apes. In the act of vaulting, the athlete grasps his pole, for example, with an exact grip that no ape can quite match. Yet such differences are secondary by comparison with the overriding difference, which is that the athlete is an adult whose behaviour is not driven by his immediate environment, as animal actions are. In themselves, his actions make no practical sense at all; they are an exercise that is not directed to the present. The athlete's mind is fixed ahead of him, building up his skill; and he vaults in imagination into the future.

Poised for that leap, the pole-vaulter is a capsule of human abilities: the grasp of the hand, the arch of the foot, the muscles of the shoulder and pelvis – the pole itself, in which energy is stored and released like a bow firing an arrow. The radical character in that complex is the sense of foresight, that is, the ability to fix an objective ahead and rigorously hold his attention on it. The athlete's performance unfolds a continued plan, from one extreme to the other, it is the invention of the pole, the concentration of the mind at the moment before leaping, which give it the stamp of humanity.

The head is more than a symbolic image of man; it is the seat of foresight and, in that respect, the spring which drives cultural evolution. Therefore if I am to take the ascent of man back to its beginnings in the animal, it is the evolution of the head and the skull that has to be traced. Unhappily, over the fifty million years or so to be talked about, there are only six or seven essentially distinct skulls which we can identify as stages in that evolution. Buried in the fossil record there must be many other intermediate steps, some of which will be found; but meanwhile we must conjecture what happened, approximately, by interpolating between the known skulls. The best way to calculate these geometrical transitions from skull to skull is on a computer; so that, in order to trace the continuity, I present them on a computer with a visual display which will lead from one to the next.

Begin fifty million years ago with a small tree-dwelling creature, a lemur; the name, appropriately, is that of the Roman spirits of the dead. The fossil skull belongs to the lemur family *Adapis*, and was found in chalky deposits outside Paris. When the skull is turned upside down, you can see the *foramen magnum* far at the back – this is a creature that hung, not held, its head on the spine. The likelihood is that it ate insects as well as fruits, and it has more than the thirty-two teeth that man and most primates now have.

The fossil lemur has some essential marks of the primates, that is, the family of monkey, ape and man. From remains of the whole skeleton we know that it has finger nails, not claws. It has a thumb that can be opposed at least in part to the hand. And it has in the skull two features that really mark the way to the beginning of man. The snout is short; the eyes are large and widely spaced. That means that there has been selection against the sense of smell and in favour of the sense of vision. The eye-sockets are still rather sideways in the skull, on either side of the snout; but compared with the eyes of earlier insect eaters, the lemur's have begun to move to the front and to give some stereoscopic vision. These are small signs of an evolutionary development towards the sophisticated structure of the human face; and yet, from that, man begins.

That was fifty million years ago, in very round figures. In the next twenty million years, the line that leads to the monkeys branches away from the main line to the apes and man. The next creature on the main line, thirty million years ago, was the fossil skull found in the Fayurn in Egypt, and named *Aegyptopithecus*. He has a shorter snout than the lemur, his teeth are ape-like, and he is larger – yet still lives in the trees. But from now on the ancestors of the apes and man spent part of their time on the ground.

Another ten million years on take us to twenty million years ago, when there were what we should now call anthropoid apes in East Africa, Europe and Asia. A classical find made by Louis Leakey goes by the dignified name of *Proconsul*, and there was at least one other widespread genus, *Dryopithecus*. (The name *Proconsul* is a piece of anthropological wit; it was coined to suggest that he was an ancestor of a famous chimpanzee at the London Zoo in 1931 whose nickname was Consul.) The brain is markedly larger, the eyes are now fully forward in stereoscopic vision. These developments tell us how the main ape-and-man line was moving. But if, as is possible, it had already branched again, then so far as man is concerned, alas, this creature is on the branch line – the ape line. The teeth show us that he is an ape, because the way in which the jaw is locked by the big canines is not man-like.

It is the change in the teeth that signals the separation of the line that leads to man, when it comes. The first harbinger that we have is *Ramapithecus*, found in Kenya and in India. This creature is fourteen million years old, and we only have pieces of the jaw. But it is clear that the teeth are level and more human. The great canines of the anthropoid apes are gone, the face is much flatter, and we are evidently near a branching of the evolutionary tree; some anthropologists would boldly put *Ramapithecus* among the hominids.

There is now a blank in the fossil record of five to ten million years. Inevitably, the blank hides the most intriguing part of the story, when the hominid line to man is firmly separated from the line to the modern apes. But we have found no unequivocal record of that, yet. Then, perhaps five million years ago, we come certainly to the relatives of man.

A cousin of man, not in the direct line to us, is a heavily-built *Australopithecus* who is a vegetarian. *Australopithecus robustus* is manlike and his line does not lead elsewhere; it has simply become extinct. The evidence that he lived on plants is again in his teeth, and it is quite direct: the teeth that survive are pitted by the fine grit that he picked up with the roots that he ate.

His cousin on the line to man is lighter – visibly so in the jaw – and is probably a meat-eater. He is the nearest thing we have to what used to be called the 'missing link': *Australopithecus africanus*, one of a number of fossil skulls found at Sterkfontein in the Transvaal and elsewhere in Africa, a fully grown female. The Taung child, with which I began, would have grown up to be like her; fully erect, walking, and with a largish brain weighing between a pound and a pound and a half. That is the size of the brain of a big ape now; but of course this was a small creature standing only four feet high. Indeed, recent finds by Richard Leakey suggest that by two million years ago the brain was larger even than that.

And with that larger brain the ancestors of man made two major inventions, for one of which we have visible evidence and for the other inferential evidence. First, the visible invention. Two million years ago *Australopithecus* made rudimentary stone tools where a simple blow has put an edge on the pebble. And for the next million years, man in his further evolution did not change this type of tool. He had made the fundamental invention, the purposeful act which prepares and stores a pebble for later use. By that lunge of skill and foresight, a symbolic act of discovery of the future, he had released the brake which the environment imposes on all other creatures. The steady use of the same tool for so long shows the strength of the invention. It was held in a simple way, by pressing its thick end against the palm of the hand in a power-grip. (The ancestors of man had a short thumb, and therefore could not manipulate very delicately, but could use the power-grip.) And, of course, it is a meat-eater's tool almost certainly, to strike and to cut.

The other invention is social, and we infer it by more subtle arithmetic. Skulls and skeletons of *Australopithecus* that have now been found in largish numbers show that most of them died before the age of twenty. That means that there must have been many orphans. For *Australopithecus* surely had a long childhood, as all the primates do; at the age of ten, say, the survivors were still children. Therefore there must have been a social organisation in which children were looked after and (as it were) adopted, were made part of the community, and so in some general sense were educated. That is a great step towards cultural evolution.

At what point can we say that the precursors of man become man himself? That is a delicate question, because such changes do not take place overnight. It would be foolish to try and make them seem more sudden than they really were – to fix the transition too sharply or to argue about names. Two million years ago we were not yet men. One million years ago we were, because by one million years ago a creature appears who can be called *Homo* – *Homo erectus*. He spreads far beyond Africa. The classical find of *Homo erectus* was in fact made in China. He is Peking man, about four hundred thousand years old, and he is the first creature that certainly used fire.

The changes in *Homo erectus* that have led to us are substantial over a million years, but they seem gradual by comparison with those that went before. The successor that we know best was first found in Germany in the last century: another classic fossil skull, he is Neanderthal man. He already has a three-pound brain, as large as modern man. Probably some lines of Neanderthal man died out; but it seems likely that a line in the Middle East went on directly to us, *Homo sapiens*.

Somewhere in that last million years or so, man made a change in the quality of his tools – which presumably points to some biological refinement in the hand during this period, and especially in the brain centres that control the hand. The more sophisticated creature (biologically and culturally) of the last half million years or so could do better than copy the ancient stone choppers that went back to *Australopithecus*. He made tools which require much finer manipulation in the making and, of course, in the use.

The development of such refined skills as this and the use of fire is not an isolated phenomenon. On the contrary, we must always remember that the real content of evolution (biological as well as cultural) is the elaboration of new behaviour. It is only because behaviour leaves no fossils that we are forced to search for it in bones and teeth. Bones and teeth are not interesting in themselves, even to the creature to whom they belong; they serve him as equipment for action – and they are interesting to us because, as equipment, they reveal his actions, and changes in equipment reveal changes in behaviour and skill.

For this reason, changes in man during his evolution did not take place piecemeal. He was not put together from the cranium of one primate and the jaw of another – that misconception is too naive to be real, and only makes a fake like the Piltdown skull. Any animal, and man especially, is a highly integrated structure, all the parts of which must change together as his behaviour changes. The evolution of the brain, of the hand, of the eyes, of the feet, the teeth, the whole human frame, made a mosaic of special gifts – and in a sense these chapters are each an essay on some special gift of man. They have made him what he is, faster in evolution, and richer and more flexible in behaviour, than any other animal. Unlike the creatures (some insects, for instance) that have been unchanged for five, ten, even fifty million years, he has changed over this time-scale out of all recognition. Man is not the most majestic of the creatures. Long before the mammals even, the dinosaurs were far more splendid. But he has what no other animal possesses, a jig-saw of faculties which alone, over three thousand million years of life, make him creative. Every animal leaves traces of what it was; man alone leaves traces of what he created.

Change in diet is important in a changing species over a time as long as fifty million years. The earliest creatures in the sequence leading to man were nimble-eyed and delicate-fingered insect and fruit eaters like the lemurs. Early apes and hominids, from *Aegyptopithecus* and *Proconsul* to the heavy *Australopithecus*, are thought to have spent their days rummaging mainly for vegetarian foods. But the light *Australopithecus* broke the ancient primate habit of vegetarianism.

The change from a vegetarian to an omnivorous diet, once made, persisted in *Homo erectus*, Neanderthal man and *Homo sapiens*. From the ancestral light *Australopithecus* onwards, the family of man ate some meat: small animals at first, larger ones later. Meat is a more concentrated protein than plant, and eating meat cuts down the bulk and the time spent in eating by two-thirds. The consequences for the evolution of man were far-reaching. He had more time free, and could spend it in more indirect ways, to get food from sources (such as large animals) which could not be tackled by hungry brute force. Evidently that helped to promote (by natural selection) the tendency of all primates to interpose an internal delay in the brain between stimulus and response, until it developed into the full human ability to postpone the gratification of desire.

But the most marked effect of an indirect strategy to enhance the food supply is, of course, to foster social action and communication. A slow creature like man can stalk, pursue and corner a large savannah animal that is adapted for flight only by co-operation. Hunting requires conscious planning and organisation by means of language, as well as special weapons. Indeed, language as we use it has something of the character of a hunting plan, in that (unlike the animals) we instruct one another in sentences which are put together from movable units. The hunt is a communal undertaking of which the climax, but only the climax, is the kill.

Hunting cannot support a growing population in one place; the limit for the savannah was not more than two people to the square mile. At that density, the total land surface of the earth could only support the present population of California, about twenty millions, and could not support the population of Great Britain. The choice for the hunters was brutal: starve or move.

They moved away over prodigious distances. By a million years ago, they were in North Africa. By seven hundred thousand years ago, or even earlier, they were in Java. By four hundred thousand years ago, they had fanned out and marched north, to China in the east and Europe in the west. These incredible spreading migrations made man, from an early time, a widely dispersed species, even though his total numbers were quite small – perhaps one million.

What is even more forbidding is that man moved north just after the climate there was turning to ice. In the great cold the ice, as it were, grew out of the ground. The northern climate had been temperate for immemorial ages – literally for several hundred million years. Yet before *Homo erectus* settled in China and northern Europe, a sequence of three separate Ice Ages began.

The first was past its fiercest when Peking man lived in caves, four hundred thousand years ago. It is no surprise to find fire used in those caves for the first time. The ice moved south and retreated three times, and the land changed each time. The icecaps at their largest contained so much of the earth's water that the level of the sea fell four hundred feet. After the second Ice Age, over two hundred thousand years ago, Neanderthal man with his big brain appears, and he became important in the last Ice Age.

The cultures of man that we recognise best began to form in the most recent Ice Age, within the last hundred or even fifty thousand years. That is when we find the elaborate tools that point to sophisticated forms of hunting: the spear-thrower, for example, and the baton that may be a straightening tool; the fully barbed harpoon; and, of course, the flint master tools that were needed to make the hunting tools.

It is clear that then, as now, inventions may be rare but they spread fast through a culture. For example, the Magdalenian hunters of southern Europe fifteen thousand years ago invented the harpoon. In the early period of the invention, the Magdalenian harpoons were unbarbed; then they were barbed with a single row of fish hooks; and at the end of the period, when the flowering of cave art took place, they were fully barbed with a double row of hooks. The Magdalenian hunters decorated their bone tools, and they can be pinned to precise periods in time and to exact geographical locations by the refinement of style which they carry. They are, in a true sense, fossils that recount the cultural evolution of man in an orderly progression.

Man survived the fierce test of the Ice Ages because he had the flexibility of mind to recognise inventions and to turn them into community property. Evidently the Ice Ages worked a profound change in the way man could live. They forced him to depend less on plants and more on animals. The rigours of hunting on the edge of the ice also changed the strategy of hunting. It became less attractive to stalk single animals, however large. The better alternative was to follow herds and not to lose them – to learn to anticipate and in the end to adopt their habits, including their wandering migrations. This is a peculiar adaptation – the transhumance mode of life on the move. It has some of the earlier qualities of hunting, because it is a pursuit; the place and the pace

Fossils that recount the cultural evolution of man in an orderly progression.

Rock painting of a reindeer hunt, Los Caballos Shelter, Valtorta Gorge, Castellon, Eastern Spain. The invention of the bow and arrow came at the end of the last Ice Age.

are set by the food animal. And it has some of the later qualities of herding, because the animal is tended and, as it were, stored as a mobile reservoir of food.

The transhumance way of life is itself a cultural fossil now, and has barely survived. The only people that still live in this way are the Lapps in the extreme north of Scandinavia, who follow the reindeer as they did during the Ice Age. The ancestors of the Lapps may have come north from the Franco-Cantabrian cave area of the Pyrenees in the wake of the reindeer as the last icecaps retreated from southern Europe twelve thousand years ago. There are thirty thousand people and three hundred thousand reindeer, and their way of life is coming to an end even now. The herds go on their own migration across the fiords from one icy pasture of lichen to another, and the Lapps go with them. But the Lapps are not herdsmen; they do not control the reindeer, they have not domesticated it. They simply move where the herds move.

Even though the reindeer herds are in effect still wild, the Lapps have some of the traditional inventions for controlling single animals that other cultures also discovered: for example, they make some males manageable as draught animals by castrating them. It is a strange relationship. The Lapps are entirely dependent on the reindeer – they eat the meat, a pound a head each every day, they use the sinews and fur and hides and bones, they drink the milk, they even use the antlers. And yet the Lapps are freer than the reindeer, because their mode of life is a cultural adaptation and not a biological one. The adaptation that the Lapps have made, the transhumance life on the move in a landscape of ice, is a choice that they can change; it is not irreversible, as biological mutations are. For a biological adaptation is an inborn form of behaviour; but a culture is a learned form of behaviour – a communally preferred form, which (like other inventions) has been adopted by a whole society.

There lies the fundamental difference between a cultural adaptation and a biological one; and both can be demonstrated in the Lapps. Making a shelter from reindeer hides is an adaptation that the Lapps can change tomorrow – most of them are doing so now. By contrast the Lapps, or human lines ancestral to them, have also undergone a certain amount of biological adaptation. The biological adaptations in *Homo sapiens* are not large; we are a rather homogeneous species, because we spread so fast over the world from a single centre. Nevertheless biological differences do exist between groups of men, as we all know. We call them racial differences, by which we mean exactly that they cannot be changed by a change of habit or habitat. You cannot change the colour of your skin. Why are the Lapps white? Man began with a dark skin; the sunlight makes vitamin D in his skin, and if he had been white in Africa, it would make too much. But in the north, man needs to let in all the sunlight there is to make enough vitamin D, and natural selection therefore favoured those with whiter skins.

The biological differences between different communities are on this modest scale. The Lapps have not lived by biological adaptation but by invention: by the imaginative use of the reindeer's habits and all its products, by turning it into a draught animal, by artefacts and the sledge. Surviving in the ice did not depend on skin colour; the Lapps have survived, man survived the Ice Ages, by the master invention of all – fire.

Fire is the symbol of the hearth, and from the time *Homo sapiens* began to leave the mark of his hand thirty thousand years ago, the hearth was the cave. For at least a million years man, in some recognisable form, lived as a forager and a hunter. We have almost no monuments of that immense period of prehistory, so much longer than any history that we record. Only at the end of that

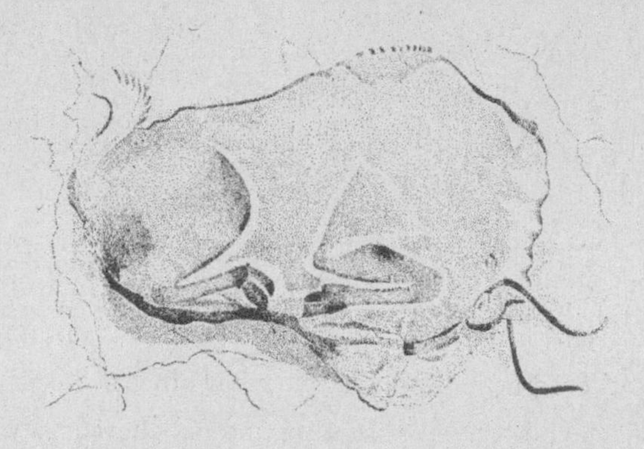

We find in caves like Altamira the record of what dominated the mind of man the hunter. I think that the power we see expressed here for the first time is the power of anticipation: the forward-looking imagination.
Recumbent bison.

time, on the edge of the European ice-sheet, we find in caves like Altamira (and elsewhere in Spain and southern France) the record of what dominated the mind of man the hunter. There we see what made his world and preoccupied him. The cave paintings, which are about twenty thousand years old, fix for ever the universal base of his culture then, the hunter's knowledge of the animal that he lived by and stalked.

One begins by thinking it odd that an art as vivid as the cave paintings should be, comparatively, so young and so rare. Why are there not more monuments to man's visual imagination, as there are to his invention? And yet when we reflect, what is remarkable is not that there are so few monuments, but that there are any at all. Man is a puny, slow, awkward, unarmed animal – he had to invent a pebble, a flint, a knife, a spear. But why to these scientific inventions, which were essential to his survival, did he from an early time add those arts that now astonish us: decorations with animal shapes? Why, above all, did he come to caves like this, live in them, and then make paintings of animals not where he lived but in places that were dark, secret, remote, hidden, inaccessible?

The obvious thing to say is that in these places the animal was magical. No doubt that is right; but magic is only a word, not an answer. In itself, magic is a word which explains nothing. It says that man believed he had power, but what power? We still want to know what the power was that the hunters believed they got from the paintings.

Here I can only give you my personal view. I think that the power that we see expressed here for the first time is the power of anticipation: the forward-looking imagination. In these paintings the hunter was made familiar with dangers which he knew he had to face but to which he had not yet come. When the hunter was brought here into the secret dark and the light was suddenly flashed on the pictures, he saw the bison as he would have to face him, he saw the running deer, he saw the turning boar. And he felt alone with them as he would in the hunt. The moment of fear was made present to him; his spear-arm flexed with an experience which he would have and which he needed not to be afraid of. The painter had frozen the moment of fear, and the hunter entered it through the painting as if through an air-lock.

For us, the cave paintings re-create the hunter's way of life as a glimpse of history; we look through them into the past. But for the hunter, I suggest, they were a peep-hole into the future; he looked ahead. In either direction, the cave paintings act as a kind of telescope tube of the imagination: they direct the mind from what is seen to what can be inferred or conjectured. Indeed, this is so in the very action of painting; for all its superb observation, the flat picture only means something to the eye because the mind fills it out with roundness and movement, a reality by inference, which is not actually seen but is imagined.

Art and science are both uniquely human actions, outside the range of anything that an animal can do. And here we see that they derive from the same human faculty: the ability to visualise the future, to foresee what may happen and plan to anticipate it, and to represent it to ourselves in images that we project and move about inside our head, or in a square of light on the dark wall of a cave or a television screen.

We also look here through the telescope of the imagination; the imagination is a telescope in time, we are looking back at the experience of the past. The men who made these paintings, the men who were present, looked through that telescope forward. They looked along the ascent of man because what we call cultural evolution is essentially a constant growing and widening of the human imagination.

The men who made the weapons and the men who made the paintings were doing the same thing – anticipating a future as only man can do, inferring what is to come from what is here. There are many gifts that are unique in man; but at the centre of them all, the root from which all knowledge grows, lies the ability to draw conclusions from what we see to what we do not see, to move our minds through space and time, and to recognise ourselves in the past on the steps to the present. All over these caves the print of the hand says: 'This is my mark. This is man.'

CHAPTER TWO

THE HARVEST OF THE SEASONS

The history of man is divided very unequally. First there is his biological evolution: all the steps that separate us from our ape ancestors. Those occupied some millions of years. And then there is his cultural history: the long swell of civilisation that separates us from the few surviving hunting tribes of Africa, or from the food-gatherers of Australia. And all that second, cultural gap is in fact crowded into a few thousand years. It goes back only about twelve thousand years – something over ten thousand years, but much less than twenty thousand. From now on I shall only be talking about those last twelve thousand years which contain almost the whole ascent of man as we think of him now. Yet the difference between the two numbers, that is, between the biological time-scale and the cultural, is so great that I cannot leave it without a backward glance.

It took at least two million years for man to change from the little dark creature with the stone in his hand, *Australopithecus* in Central Africa, to the modern form, *Homo sapiens*. That is the pace of biological evolution – even though the biological evolution of man has been faster than that of any other animal. But it has taken much less than twenty thousand years for *Homo sapiens* to become the creatures that you and I aspire to be: artists and scientists, city builders and planners for the future, readers and travellers, eager explorers of natural fact and human emotion, immensely richer in experience and bolder in imagination than any of our ancestors. That is the pace of cultural evolution; once it takes off, it goes as the ratio of those two numbers goes, at least a hundred times faster than biological evolution.

Once it takes off: that is the crucial phrase. Why did the cultural changes that have made man master of the earth begin so recently? Twenty thousand years ago man in all parts of the world that he had reached was a forager and a hunter, whose most advanced technique was to attach himself to a moving herd as the Lapps still do. By ten thousand years ago that had changed, and he had begun in some places to domesticate some animals and to cultivate some plants; and that is the change from which civilisation took off. It is extraordinary to think that only in the last twelve thousand years has civilisation, as we understand it, taken off. There must have been an extraordinary explosion about 10,000 BC – and there was. But it was a quiet explosion. It was the end of the last Ice Age.

We can catch the look and, as it were, the smell of the change in some glacial landscape. Spring in Iceland replays itself every year, but it once played itself over Europe and Asia when the ice retreated. And man, who had come through incredible hardships, had wandered up from Africa over the last million years, had battled through the Ice Ages, suddenly found the ground flowering and the animals surrounding him, and moved into a different kind of life.

It is usually called the 'agricultural revolution'. But I think of it as something much wider, the biological revolution. There was intertwined in it the cultivation of plants and the domestication of animals in a kind of leap-frog. And under this ran the crucial realisation that man dominates his environment in its most important aspect, not physically but at the level of living things – plants and animals. With that there comes an equally powerful social revolution. Because now it became possible – more than that, it became necessary – for man to settle. And this creature that had roamed and marched for a million years had to make the crucial decision: whether he would cease to be a nomad and become a villager. We have an anthropological record of the struggle of conscience of a people who make this decision: the record is the Bible, the Old Testament. I believe that civilisation rests on that decision. As for people who never made it, there are few survivors. There are some nomad tribes who still go through these vast transhumance journeys from one grazing ground to another: the Bakhtiari in Persia, for example. And you have actually to travel with them and live with them to understand that civilisation can never grow up on the move.

Everything in nomad life is immemorial. The Bakhtiari have always travelled alone, quite unseen. Like other nomads, they think of themselves as a family, the sons of a single founding father. (In the same way the Jews used to call themselves the children of Israel or Jacob.) The Bakhtiari take their name from a legendary herdsman of Mongol times, Bakhtyar. The legend of their own origin that they tell of him begins,

> And the father of our people, the hill-man, Bakhtyar, came out of the fastness of the southern mountains in ancient times. His seed were as numerous as the rocks on the mountains, and his people prospered.

The biblical echo sounds again and again as the story goes on. The patriarch Jacob had two wives, and had worked as a herdsman for seven years for each of them. Compare the patriarch of the Bakhtiari:

> The first wife of Bakhtyar had seven sons, fathers of the seven brother lines of our people. His second wife had four sons. And our sons shall take for wives the daughters from their father's brothers' tents, lest the flocks and tents be dispersed.

As with the children of Israel, the flocks were all-important; they are not out of the mind of the storyteller (or the marriage counsellor) for a moment.

Before 10,000 BC nomad peoples used to follow the natural migration of wild herds. But sheep and goats have no natural migrations. They were first domesticated about ten thousand years ago – only the dog is an older camp follower than that. And when man domesticated them, he took on the responsibility of nature; the nomad must lead the helpless herd.

The role of women in nomad tribes is narrowly defined. Above all, the function of women is to produce men-children; too many she-children are an immediate misfortune, because in the long run they threaten disaster. Apart from that, their duties lie in preparing food and clothes. For example, the women among the Bakhtiari bake bread – in the biblical manner, in unleavened cakes on hot stones. But the girls and the women wait to eat until the men have eaten. Like the men, the lives of the women centre on the flock. They milk the herd, and they make a clotted yoghourt from the milk by churning it in a goatskin bag on a primitive wooden frame. They have only the simple technology that can be carried on daily journeys from place to place. The simplicity is not romantic; it is a matter of survival. Everything must be light enough to be carried, to be set up every evening and to be packed away again every morning. When the women spin wool with their simple, ancient devices, it is for immediate use, to make the repairs that are essential on the journey – no more.

It is not possible in the nomad life to make things that will not be needed for several weeks. They could not be carried. And in fact the Bakhtiari do not know how to make them. If they need metal pots, they barter them from settled peoples or from a caste of gipsy workers who specialise in metals. A nail, a stirrup, a toy, or a child's bell is something that is traded from outside the tribe. The Bakhtiari life is too narrow to have time or skill for specialisation. There is no room for innovation, because there is not time, on the move, between evening and morning, coming and going all their lives, to develop a new device or a new thought – not even a new tune. The only habits that survive are the old habits. The only ambition of the son is to be like the father.

It is a life without features. Every night is the end of a day like the last, and every morning will be the beginning of a journey like the day before. When the day breaks, there is one question in everyone's mind: Can the flock be got over the next high pass? One day on the journey, the highest pass of all must be crossed. This is the pass Zadeku, twelve thousand feet high on the Zagros, which the flock must somehow struggle through or skirt in its upper reaches. For the tribe must move on, the herdsman must find new pastures every day, because at these heights grazing is exhausted in a single day.

Every year the Bakhtiari cross six ranges of mountains on the outward journey (and cross them again to come back). They march through snow and the spring flood water. And in only one respect has their life advanced beyond that of ten thousand years ago. The nomads of that time had to travel on foot and carry their own packs. The Bakhtiari have pack-animals – horses, donkeys, mules – which have only been domesticated since that time. Nothing else in their lives is new. And nothing is memorable. Nomads have no memorials, even to the dead. (Where is Bakhtyar, where was Jacob buried?) The only mounds that they build are to mark the way at such places as the Pass of the Women, treacherous but easier for the animals than the high pass.

The spring migration of the Bakhtiari is a heroic adventure; and yet the Bakhtiari are not so much heroic as stoic. They are resigned because the adventure leads nowhere. The summer pastures themselves will only be a stopping place – unlike the children of Israel, for them there is no promised land. The head of the family has worked seven years, as Jacob did, to build a flock of fifty sheep and goats. He expects to lose ten of them in the migration if things go well. If they go badly, he may lose twenty out of that fifty. Those are the odds of the nomad life, year in and year out. And beyond that, at the end of the journey, there will still be nothing except an immense, traditional resignation.

Who knows, in any one year, whether the old when they have crossed the passes will be able to face the final test: the crossing of the Bazuft River? Three months of melt-water have swollen the river. The tribesmen, the women, the pack animals and the flocks are all exhausted. It will take a day to manhandle the flocks across the river. But this, here, now is the testing day. Today is the day on which the young become men, because the survival of the herd and the family depends on their strength. Crossing the Bazuft River is like crossing the Jordan; it is the baptism to manhood. For the young man, life for a moment comes alive now. And for the old – for the old, it dies.

What happens to the old when they cannot cross the last river? Nothing. They stay behind to die. Only the dog is puzzled to see a man abandoned. The man accepts the nomad custom; he has come to the end of his journey, and there is no place at the end.

The largest single step in the ascent of man is the change from nomad to village agriculture. What made that possible? An act of will by men, surely; but with that, a strange and secret act of nature. In the burst of new vegetation at the end of the Ice Age, a hybrid wheat appeared in the Middle East. It happened in many places: a typical one is the ancient oasis of Jericho.

Jericho is older than agriculture. The first people who came here and settled by the spring in this otherwise desolate ground were people who harvested wheat, but did not yet know how to plant it. We know this because they made tools for the wild harvest, and that is an extraordinary piece of foresight. They made sickles out of flint which have survived; John Garstang found them when he was digging here in the 1930s. The ancient sickle edge would have been set in a piece of gazelle horn, or bone.

There no longer survives, up on the hill or tel and its slopes, the kind of wild wheat that the earliest inhabitants harvested. But the grasses that are still here must look very like the wheat that they found, that they gathered for the first time by the fistful, and cut with that sawing motion of the sickle that reapers have used for all the ten thousand years since then. That was the Natufian pre-agricultural civilisation. And, of course, it could not last. It was on the brink of becoming agriculture. And that was the next thing that happened on the Jericho tel.

The turning-point to the spread of agriculture in the Old World was almost certainly the occurrence of two forms of wheat with a large, full head of seeds. Before 8000 BC wheat was not the luxuriant plant it is today; it was merely one of many wild grasses that spread throughout the Middle Fast. By some genetic accident, the wild wheat crossed with a natural goat grass and formed a fertile hybrid. That accident must have happened many times in the springing vegetation that came up after the last Ice Age. In terms of

(*Opposite*) Jericho is monumental, older than the Bible, layer upon layer of history, a city.

From the Jericho site: plaster-decorated skull inset with cowrie shells.

The tower at Jericho tel. Its masonry is flint-worked and pre-7000 BC. *The modern grid covers the hollow shaft inside the tower.*

the genetic machinery that directs growth, it combined the fourteen chromosomes of wild wheat with the fourteen chromosomes of goat grass, and produced Emmer with twenty-eight chromosomes. That is what makes Emmer so much plumper. The hybrid was able to spread naturally, because its seeds are attached to the husk in such a way that they scatter in the wind.

For such a hybrid to be fertile is rare but not unique among plants. But now the story of the rich plant life that followed the Ice Ages becomes more surprising. There was a second genetic accident, which may have come about because Emmer was already cultivated. Emmer crossed with another natural goat grass and produced a still larger hybrid with forty-two chromosomes, which is bread wheat. That was improbable enough in itself, and we know now that bread wheat would not have been fertile but for a specific genetic mutation on one chromosome.

Yet there is something even stranger. Now we have a beautiful ear of wheat, but one which will never spread in the wind because the ear is too tight to break up. And if I do break it up, why, then the chaff flies off and every grain falls exactly where it grew. Let me remind you, that is quite different from the wild wheats or from the first, primitive hybrid, Emmer. In those primitive forms the ear is much more open, and if the ear breaks up then you get quite a different effect – you get grains which will fly in the wind. The bread wheats have lost that ability. Suddenly, man and the plant have come together. Man has a wheat that he lives by, but the wheat also thinks that man was made for him because only so can it be propagated. For the bread wheats can only multiply with help; man must harvest the ears and scatter their seeds; and the life of each, man and the plant, depends on the other. It is a true fairy tale of genetics, as if the coming of civilisation had been blessed in advance by the spirit of the abbot Gregor Mendel.

A happy conjunction of natural and human events created agriculture. In the Old World that happened about ten thousand years ago, and it happened in the Fertile Crescent of the Middle East. But it surely happened more than once. Almost certainly agriculture was invented again and independently in the New World – or so we believe on the evidence we now have that maize needed man like wheat. As for the Middle East, agriculture was spread here and there over its hilly slopes, of which the climb from the Dead Sea to Judea, the hinterland of Jericho, is at best a characteristic piece and no more. In a literal sense, agriculture is likely to have had several beginnings in the Fertile Crescent, some of them before Jericho.

Yet Jericho has several features which make it historically unique and give it a symbolic status of its own. Unlike the forgotten villages elsewhere, it is monumental, older than the Bible, layer upon layer of history, a city. The ancient sweet-water city of Jericho was an oasis on the edge of the desert whose spring has been running from prehistoric times right into the modern city today. Here wheat and water came together and, in that sense, here man began civilisation. Here, too, the bedouin came with their dark muffled faces out of the desert, looking jealously at the new way of life. That is why Joshua brought the tribes of Israel here on their way to the Promised Land – because wheat and water, they make civilisation: they make the promise of a land flowing with milk and honey. Wheat and water turned that barren hillside into the oldest city of the world.

All at once at that time Jericho is transformed. People come and soon become the envy of their neighbours, so that they have to fortify Jericho, turn it into a walled city, and build a stupendous tower, nine thousand years ago. The tower is thirty feet across at the base and, to correspond, almost thirty feet in depth. And climbing up beside it the excavation reveals layer upon layer of past civilisation: the early pre-pottery men, the next pre-pottery men, the coming of pottery seven thousand years ago; early copper, early bronze, middle bronze. Each of these civilisations came, conquered Jericho, buried it, and built itself up; so that the tower lies not so much under forty-five feet of soil as under forty-five feet of past civilisations.

Jericho is a microcosm of history. There will be other sites found in coming years (there are some important new ones already) which will change our picture of the beginnings of civilisation. Yet the power of standing in this place, the vision backward along the ascent of modern man, is profound in thought and in emotion equally. When I was a young man, we all thought that mastery came from man's domination of his physical environment. Now we have learned that real mastery comes from understanding and moulding the living environment. That is how man began in the Fertile Crescent when he put his hand on plant and animal and, in learning to live with them, changed the world to his needs. When Kathleen Kenyon rediscovered the ancient tower in the 1950s, she found that it was hollow; and to me, this staircase is a sort of taproot, a peephole to the rock base of civilisation. And the rock base of civilisation is the living being, not the physical world.

By 6000 BC Jericho was a large agricultural settlement. Kathleen Kenyon estimates that it contained three thousand people, and covered eight or ten acres within the walls. The women ground the wheat with the heavy stone implements that characterise a settled community. The men shaped, patted and moulded the clay for building-bricks, some of the earliest known. The marks of the brick-makers' thumbprints are still there. Man, like the bread wheat, is now fixed in his place. A settled community also has a different relation to the dead. The inhabitants of Jericho preserved some skulls and covered them with elaborate decoration. No one knows why, unless it was a reverential action.

No one who was brought up on the Old Testament, as I was, can leave Jericho without asking two questions: Did Joshua finally destroy this city? And did the walls really come tumbling down? Those are the questions that bring people to this site and turn it into a living legend. To the first question, there is an easy answer: Yes. The tribes of Israel were fighting to get into the Fertile Crescent which runs up the Mediterranean coast, along the mountains of Anatolia, and down towards the Tigris and Euphrates. And here at Jericho was the key that locked their way up the mountains of Judea and out into the Mediterranean fertile land. This they had to conquer, and they did about 1400 BC – about three thousand three hundred to three thousand four hundred years ago. The Bible story was not written down until perhaps 700 BC; that is, the account is about two thousand six hundred years old as a written record.

But did the walls come tumbling down? We do not know. There is no archaeological evidence on this site that suggests that a set of walls one fine day really fell flat. But many sets of walls did fall, at different times. There is a Bronze Age period here where a set of walls was rebuilt at least sixteen times. Because this is earthquake country. There are tremors here still every day; there are four major quakes in a century. It is only in the last years that we have come to understand why earthquakes run along this valley. The Red Sea and the Dead Sea lie along a continuation of the Great Rift Valley of East Africa. Here two of the plates that carry the continents as they float on the denser mantle of the earth ride side by side. As they thrust past one another along this rift, the surface of the earth echoes to the shocks that well up from below. As a result, earthquakes have always erupted along the axis on which the Dead Sea lies. And in my view that is why the Bible is full of memories of natural miracles: some ancient flood, some running dry of the Red Sea, the Jordan running dry, and the walls of Jericho falling down.

The Bible is a curious history, part folklore and part record. History is, of course, written by the victors, and the Israelis, when they burst through here, became the carriers of history. The Bible is their story: the history of a people who had to stop being nomad and pastoral and had to become an agricultural tribe.

Farming and husbandry seem simple pursuits, but the Natufian sickle is a signal to show us that they do not stand still. Every stage in the domestication of plant and animal life requires inventions, which begin as technical devices and from which flow scientific principles. The basic devices of the nimble-fingered mind lie about, unregarded, in any village anywhere in the world. Their cornucopia of small and subtle artifices is as ingenious, and in a deep sense as important in the ascent of man, as any apparatus of nuclear physics: the needle, the awl, the pot, the brazier, the spade, the nail and the screw, the bellows, the string, the knot, the loom, the harness, the hook, the button, the shoe – one could name a hundred and not stop for breath. The richness comes from the interplay of inventions; a culture is a multiplier of ideas, in which each new device quickens and enlarges the power of the rest.

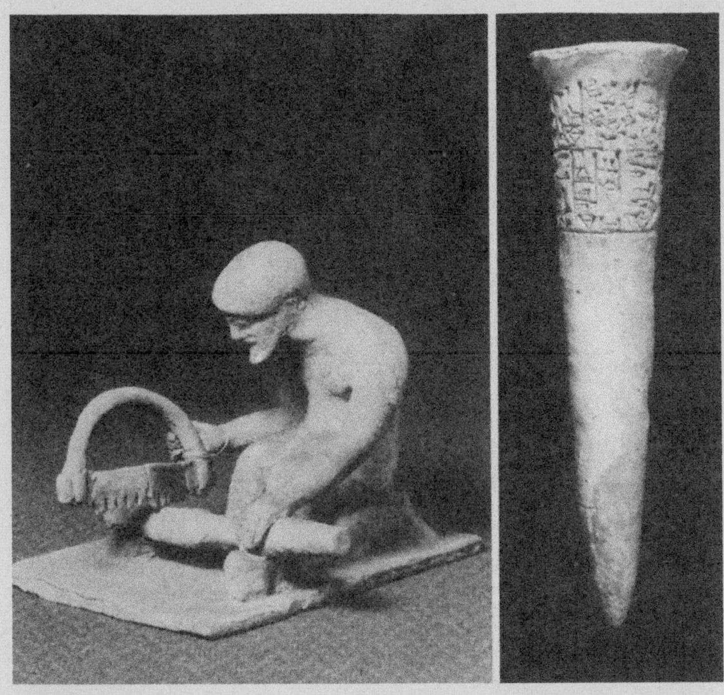

(*Opposite*) A cornucopia of small and subtle artifices as important in the ascent of man as any apparatus of nuclear physics.
Carpenter working on a piece of turned wood with a saw. Greek, 6th century BC.
Clay treaty nail, Sumerian, 2400 BC.
Baker's oven with bread cooking. Clay model. Greek Islands, 7th century BC.

Settled agriculture creates a technology from which all physics, all science takes off. We can see it in the change from the early sickle to the late. At first glance they look very much alike: the sickle of ten thousand years ago of the gatherer, and the sickle of nine thousand years ago when wheat was cultivated. But look more closely. The cultivated wheat is sawed with a serrated edge: because if you hit the wheat, then the grains will fall to the ground; but if you gently saw it, the grains will be held in the ear of corn. And sickles have been made like this ever since then – into my boyhood in the First World War, when the curved sickle with the serrated edge was still what you cut wheat with. A technology like that, physical knowledge like that, comes to us out of every part of the agricultural life so spontaneously that we feel as if the ideas discover man, rather than the other way about.

The most powerful invention in all agriculture is, of course, the plough. We think of the plough as a wedge dividing the soil. And the wedge is an important early mechanical invention. But the plough is also something much more fundamental: it is a lever which lifts the soil, and it is among the first applications of the principle of the lever. When, long afterwards, Archimedes explained the theory of the lever to the Greeks, he said that with a fulcrum for the lever he could move the earth. But thousands of years before that the ploughmen of the Middle East had been saying 'Give me a lever and I will feed the earth'.

I have remarked that agriculture was invented at least once again, much later, in America. But the plough and the wheel were not, because they depend on the draught animal. The step beyond simple agriculture in the Middle East was the domestication of draught animals. The failure to make that biological move kept the New World back at the level of the digging stick and the pack; it did not even hit on the potter's wheel.

The wheel is found for the first time before 3000 BC in what is now southern Russia. These early finds are solid wooden wheels attached to an older raft or sledge for drawing loads, which thereby is converted into a cart. From then on the wheel and axle becomes the double root from which invention grows. For example, it is turned into an instrument for grinding wheat – and using the forces of nature to do that: the animal forces first, and later the forces of wind and water. The wheel becomes a model for all motions of rotation, a norm of explanation and a heavenly symbol of more than human power in science and in art alike. The sun is a wheeled chariot, and the sky itself is a wheel, from the time that the Babylonians and the Greeks mapped the turning of the starry heavens. In modern science natural motion (that is, undisturbed motion) goes in a straight line; but for Greek science, the form of motion that seemed natural (that is, inherent in nature) and in fact perfect was motion in a circle.

About the time that Joshua stormed Jericho, say 1400 BC, the mechanical engineers of Sumer and Assyria turned the wheel into a pulley to draw water. At the same time they designed large-scale irrigation systems. The vertical maintenance shafts still survive like punctuation marks across the Persian landscape. They go down three hundred feet to the qanats or underground canals that make up the system, at a level where the natural water is safe from evaporation. Three thousand years after they were made, the village women of Khuzistan still draw their water ration from the qanats to carry on the everyday chores of ancient communities.

The bow-lathe is one of the classical schemes for turning linear into rotary motion.

Mid 19th-century carpenters at work with a bow-lathe, Central India.

The qanats are a late construction of a city civilisation, and they imply the existence by then of laws to govern water rights and land tenure and other social relations. In an agricultural community (the large-scale peasant farming of Sinner, for instance) the rule of law has a different character from the nomad law that governs the theft of a goat or a sheep. Now the social structure is bound up with the regulation of matters that affect the community as a whole: access to land, the upkeep and control of water rights, the right to use, turn and turn about, the precious constructions on which the harvest of the seasons depends.

By now the village artisan has become an inventor in his own right. He combines the basic mechanical principles in sophisticated tools which are, in effect, early machines. They are traditional in the Middle East: the bow-lathe, for example, which is one of the classical schemes for turning linear into rotary motion. Here the scheme depends, ingeniously, on winding a string round a drum and fastening the ends of the string to the two ends of a sort of violin bow. The piece of wood to be worked is fixed to the drum; it is turned by moving the bow to and fro, so that the string rotates the drum that holds the piece of wood, which is scored by a chisel. The combination is several thousand years old, but I saw it used by gipsies making chair-legs in a wood in England in 1945.

A machine is a device for tapping the power in nature. That is true from the simplest spindle that the Bakhtiari women carry, all the way to the historic first nuclear reactor and all its busy progeny. Yet as the machine has tapped larger sources of power, it has come more and more to outdistance its natural use. How is it that the machine in its modern form now seems to us a threat?

The question as it strikes us hinges on the scale of power that the machine can develop. We can put it in the form of alternatives: Is the power within the scale of the work for which the machine was devised, or is it so disproportionate that it can dominate the user and distort the use? The question therefore reaches far back; it begins when man first harnessed a power greater than his own, the power of animals. Every machine is a kind of draught animal – even the nuclear reactor. It increases the surplus that man has won from nature since the beginning of agriculture. And therefore every machine re-enacts the original dilemma: does it deliver energy in response to the demand of its specific use, or is it a maverick source of energy beyond the limits of constructive use? The conflict in the scale of power goes back all the way to that formative time in human history.

Agriculture is one part of the biological revolution; the domestication and harnessing of village animals is the other. The sequence of domestication is orderly. First comes the dog, perhaps even before 10,000 BC. Then come food animals, beginning with goats and sheep.

And then come draught animals such as the onager, a kind of wild ass. The animals add a surplus much larger than they consume. But that is true only so long as the animals remain modestly in their proper station, as servants of agriculture.

It is unexpected that the domestic animal should turn out exactly to contain within itself, from then on, the threat to the surplus of grain by which the settled community lives and survives. Most unexpected, because after all it is the ox, the ass, as a draught animal that has helped to create this surplus. (The Old Testament carefully urges that they be treated well; for instance, it forbids the farmer to yoke an ox and an ass to the plough together, since they work in different ways.) But round about five thousand years ago, a new draught animal appears – the horse. And that is out of all proportion faster, stronger, more dominant than any previous animal. And from now on that becomes the threat to the village surplus.

The horse had begun by drawing wheeled carts, like the ox but rather grander, drawing chariots in the processions of kings. And then, somewhere around 2000 BC, men discovered how to ride it. The idea must have been as startling in its day as the invention of the flying machine. For one thing, it required a bigger, stronger horse – the horse was originally quite a small animal and, like the llama of South America, could not carry a man for long. Riding as a serious use for the horse therefore begins in the nomad tribes that bred horses. They were men out of Central Asia, Persia, Afghanistan and beyond; in the west they were simply called Scythians, as a collective name for a new and frightening creature, a phenomenon of nature.

For the rider visibly is more than a man: he is head-high above others, and he moves with bewildering power so that he bestrides the living world. When the plants and the animals of the village had been tamed for human use, mounting the horse was a more than human gesture, the symbolic act of dominance over the total creation. We know that this is so from the awe and fear that the horse created again in historical times, when the mounted Spaniards overwhelmed the armies of Peru (who had never seen a horse) in 1532. So, long before, the Scythians were a terror that swept over the countries that did not know the technique of riding. The Greeks when they saw the Scythian riders believed the horse and the rider to be one; that is how they invented the legend of the centaur. Indeed, that other half-human hybrid of the Greek imagination, the satyr, was originally not part goat but part horse; so deep was the unease that the rushing creature from the east evoked.

We cannot hope to recapture today the terror that the mounted horse struck into the Middle East and Eastern Europe when it first appeared. That is because there is a difference of scale which I can only compare with the arrival of tanks in Poland in 1939, sweeping all before them. I believe that the importance of the horse in European history has always been underrated. In a sense, warfare was created by the horse, as a nomad activity. That is what the Huns brought, that is what the Phrygians brought, that is what finally the Mongols brought, and brought to a climax under Genghis Khan much later. In particular, the mobile hordes transformed the organisation of battle. They conceived a different strategy of war – a strategy that is like a war game; how, warmakers love to play games!

The strategy of the mobile horde depends on manoeuvre, on rapid communication, and on practised tactical moves which can be strung together into different sequences of surprise. The remnants of that remain in the war games that are still played and that come from Asia, such as chess and polo. War strategy is always regarded by those who win as a kind of game. And there is played to this day in Afghanistan a game called Buz Kashi which comes from the kind of competitive riding that was carried on by the Mongols.

The Greeks when they saw the Scythian riders believed the horse and the rider to be one; that is how they invented the legend of the centaur.
Greek vase painting, c.560 BC. Centaurs and a warrior arming.

The men who play the game of Buz Kashi are professionals – that is to say, they are retainers, and they and the horses are trained and kept simply for the glory of winning. On a great occasion three hundred men from different tribes would come to compete, though that had not happened now for twenty or thirty years, until we organised it.

The players in the game of Buz Kashi do not form teams. The object of the game is not to prove one group better than another, but to find a champion. There are famous champions from the past, and they are remembered. The President who supervised this game was a champion who no longer played. The President gives his orders through a herald, who may also be a pensioner of the game, though less distinguished. Where we should expect to see a ball, there is instead a headless calf. (And that macabre plaything says something about the game, as if the riders were making sport of the farmers' livelihood.) The carcass weighs about fifty pounds and the object is to snatch it up, defending it against all challengers, and carry it off through two stages. The first stage of the game is riding off with the carcass to the fixed boundary flag and rounding the flag. After that the crucial stage is the return; as he sweeps round the flag, constantly challenged, the rider heads for home and the goal, which is a marked circle in the centre of the mêlée.

The game is going to be won by a single goal, so no quarter is given. This is not a sporting event; there is nothing in the rules about fair play. The tactics are pure Mongol, a discipline of shock. The astonishing thing in the game is what routed the armies that faced the Mongols: that what seems a wild scrimmage is in fact full of manoeuvre, and dissolves suddenly with the winner riding clear to score.

One has the sense that the crowd is much more excited, and more involved emotionally, than the players. The players, by contrast, seem committed but cold; they ride with a brilliant and brutal intensity, but they are not absorbed in playing, they are absorbed in winning. Only after the game is the winner himself carried away by the excitement. He should have asked the President to sanction the goal and, by missing that point of etiquette in this uproar, he has jeopardised the goal. It is nice to know that the goal was allowed.

The Buz Kashi is a war game. What makes it electric is the cowboy ethic: riding as an act of war. It expresses the monomaniac culture of conquest; the predator posing as a hero because he rides the whirlwind. But the whirlwind is empty. Horse or tank, Genghis Khan or Hitler or Stalin, it can only feed on the labours of other men. The nomad in his last historic role as warmaker is still an anachronism, and worse, in a world that has discovered, in the last twelve thousand years, that civilisation is made by settled people.

All through this essay there runs the conflict between the nomad and the settled way of life. So it is fitting by way of epitaph to go to that high, windy, inhospitable plateau at Sultaniyeh in Persia where ended the last attempt by the Mongol dynasty of Genghis Khan to make the nomad way of life supreme. The point is that the invention of agriculture twelve thousand years ago did not of itself establish or confirm the settled way of life. On the contrary, the domestication of animals that came with agriculture gave new vigour to nomad economics: the domestication of the sheep and the goat, for example, and then, above all, the domestication of the horse. It was the horse that gave the Mongol hordes of Genghis Khan the power and the organisation to conquer China and the Muslim states and to reach the gates of central Europe.

Genghis Khan was a nomad and the inventor of a powerful war machine – and that conjunction says something important about the origins of war in human history. Of course, it is tempting to close one's eyes to history, and instead to speculate about the roots of war in some possible animal instinct: as if, like the tiger, we still had to kill to live, or, like the robin redbreast, to defend a nesting territory. But war, organised war, is not a human instinct. It is a highly planned and co-operative form of theft. And that form of theft began ten thousand years ago when the harvesters of wheat accumulated a surplus, and the nomads rose out of the desert to rob them of what they themselves could not provide. The evidence for that we saw in the walled city of Jericho and its prehistoric tower. That is the beginning of war.

Genghis Khan and his Mongol dynasty brought that thieving way of life into our own millennium. From AD 1200 to 1300 they made almost the last attempt to establish the supremacy of the robber who produces nothing and who, in his feckless way, comes to take from the peasant (who has nowhere to flee) the surplus that agriculture accumulates.

Yet that attempt failed. And it failed because in the end there was nothing for the Mongols to do except themselves to adopt the way of life of the people that they had conquered. When they conquered the Muslims, they became Muslims. They became settlers because theft, war, is not a permanent state that can be sustained. Of course, Genghis Khan still had his bones carried about as a memorial by his armies in the field. But his grandson Kublai Khan was already a builder and settled monarch in China; you remember Coleridge's poem,

> In Xanadu did Kubla Khan
> A stately pleasure-dome decree.

The fifth of the heirs in succession to Genghis Khan was the sultan Oljeitu, who came to this forbidding plateau in Persia to build a great new capital city, Sultaniyeh. What remains is his own mausoleum which later was a model for much Muslim architecture. Oljeitu was a liberal monarch, who brought here men from all parts of the world. He himself was a Christian, at another time a Buddhist, and finally a Muslim, and he did – at this court – attempt really to establish a world court. It was the one thing that the nomad could contribute to civilisation: he gathered from the four corners of the world the cultures, mixed them together, and sent them out again to fertilise the earth.

It is the irony of the end of the bid for power by the Mongol nomads here that when Oljeitu died, he was known as Oljeitu the Builder. The fact is that agriculture and the settled way of life were established steps now in the ascent of man, and had set a new level for a form of human harmony which was to bear fruit into the far future: the organisation of the city.

CHAPTER THREE

THE GRAIN IN THE STONE

In his hand
He took the golden Compasses, prepar'd
In Gods Eternal store, to circumscribe
This Universe, and all created things:
One foot he center'd, and the other turn'd
Round through the vast profunditie obscure,
And said, thus farr extend, thus farr thy bounds,
This be thy just Circumference, O World.

Milton, *Paradise Lost*, Book VII

John Milton described and William Blake drew the shaping of the earth in a single sweeping motion by the compasses of God. But that is an excessively static picture of the processes of nature. The earth has existed for more than four thousand million years. Through all this time, it has been shaped and changed by two kinds of action. The hidden forces within the earth have buckled the strata, and lifted and shifted the land masses. And on the surface, the erosion of snow and rain and storm, of stream and ocean, of sun and wind, have carved out a natural architecture.

Man has also become an architect of his environment, but he does not command forces as powerful as those of nature. His method has been selective and probing: an intellectual approach in which action depends on understanding. I have come to trace its history in the cultures of the New World which are younger than Europe and Asia. I centred my first essay on equatorial Africa, because that is where man began, and my second essay on the Near East, because that is where civilisation began. Now it is time to remember that man reached other continents too in his long walk over the earth.

The Canyon de Chelly in Arizona is a breathless, secret valley, which has been inhabited by one Indian tribe after another almost without a break for two thousand years, since the birth of Christ; longer than any other place in North America. Sir Thomas Browne has a springing sentence: 'The Huntsmen are up in America, and they are already past their first sleep in Persia.' At the birth of Christ, the huntsmen were settling to agriculture in the Canyon de Chelly, and starting along the same steps in the ascent of man that had first been taken in the Fertile Crescent of the Middle East.

Why did civilisation begin so much later in the New World than in the Old? Evidently because man was a latecomer to the New World. He came before boats were invented, which implies that he came dry-shod over the Bering Straits when they formed a broad land-bridge during the last Ice Age. The glaciological evidence points to two possible times when men might have wandered from the easternmost promontories of the Old World beyond Siberia to the rocky wastes of western Alaska in the New. One period was between 28,000 BC and 23,000 BC, and the other between 14,000 BC and 10,000 BC. After that the flood of melt-water at the end of the last Ice Age raised the sea level again by several hundred feet and thereby turned the key on the inhabitants of the New World.

That means that man came from Asia to America not later than ten thousand years ago, and not earlier than about thirty thousand years ago. And he did not necessarily come all at once. There is evidence in archaeological finds (such as early sites and tools) that two separate streams of culture came to America. And, most telling to me, there is subtle but persuasive biological evidence that I can only interpret to mean that he came in two small, successive migrations.

The Indian tribes of North and South America do not contain all the blood groups that are found in populations elsewhere. A fascinating glimpse into their ancestry is opened by this unexpected biological quirk. For the blood groups are inherited in such a way that, over a whole population, they provide some genetic record of the past. The total absence of blood group A from a population implies, with virtual certainty, that there was no blood group A in its ancestry; and similarly with blood group B. And this is in fact the state of affairs in America. The tribes of Central and South America (in the Amazon, for example, in the Andes, and in Tierra del Fuego) belong entirely to blood group O; so do some North American tribes. Others (among them the Sioux, the Chippewa, and the Pueblo Indians) consist of blood group O mixed with ten to fifteen per cent of blood group A.

In summary, the evidence is that there is no blood group B anywhere in America, as there is in most other parts of the world.

In Central and South America, all the original Indian population is blood group O. In North America, it is of blood groups O and A. I can see no sensible way of interpreting that but to believe that a first migration of a small, related kinship group (all of blood group O) came into America, multiplied, and spread right down to the south. Then a second migration, again of small groups, this time containing either A alone or both A and O, followed them only as far as North America. The American Indians of the north, then, certainly contain some of this later migration and are, comparatively speaking, latecomers.

Agriculture in the Canyon de Chelly reflects this lateness. Although maize had long been cultivated in Central and South America, here it comes in only about the time of Christ. People are very simple, they have no houses, they live in caves. About AD 500 pottery is introduced. Pit houses are dug in the caves themselves, and covered with a roof moulded out of clay or adobe. And at that stage the Canyon is really fixed until about the year AD 1000, when the great Pueblo civilisation comes in with stone masonry.

I am making a basic separation between architecture as moulding and architecture as the assembly of parts. That seems a very simple distinction: the mud house, the stone masonry. But in fact it represents a fundamental intellectual difference, not just a technical one. And I believe it to be one of the most important steps that man has taken, wherever and whenever he did so: the distinction between the moulding action of the hand, and the splitting or analytic action of the hand.

It seems the most natural thing in the world to take some clay and mould it into a ball, a little clay figure, a cup, a pit house. At first we feel that the shape of nature has been given us by this. But, of course, it has not. This is the man-made shape. What the pot does is to reflect the cupped hand; what the pit house does is to reflect the shaping action of man. And nothing has been discovered about nature herself when man imposes these warm, rounded, feminine, artistic shapes on her. The only thing that you reflect is the shape of your own hand.

But there is another action of the human hand which is different and opposite. That is the splitting of wood or stone; for by that action the hand (armed with a tool) probes and explores beneath the surface, and thereby becomes an instrument of discovery. There is a great intellectual step forward when man splits a piece of wood, or a piece of stone, and lays bare the print that nature had put there before he split it. The Pueblo people found that step in the red sandstone cliffs that rise a thousand feet over the Arizona settlements. The tabular strata were there for the cutting; and the blocks were laid in courses along the same bedding planes in which they had lain in the cliffs of the Canyon de Chelly.

From an early time man made tools by working the stone. Sometimes the stone had a natural grain, sometimes the toolmaker created the lines of cleavage by learning how to strike the stone. It may be that the idea comes, in the first place, from splitting wood, because wood is a material with a visible structure which opens easily along the grain, but which is hard to shear across the grain. And from that simple beginning man prises open the nature of things and uncovers the laws that the structure dictates and reveals. Now the hand no longer imposes itself on the shape of things. Instead, it becomes an instrument of discovery and pleasure together, in which the tool transcends its immediate use and enters into and reveals the qualities and the forms that lie hidden in the material. Like a man cutting a crystal, we find in the form within the secret laws of nature.

The notion of discovering an underlying order in matter is man's basic concept for exploring nature. The architecture of things reveals a structure below the surface, a hidden grain which, when it is laid bare, makes it possible to take natural formations apart and assemble them in new arrangements. For me this is the step in the ascent of man at which theoretical science begins. And it is as native to the way man conceives his own communities as it is to his conception of nature.

We human beings are joined in families, the families are joined in kinship groups, the kinship groups in clans, the clans in tribes, the tribes in nations. And that sense of hierarchy, of a pyramid in which layer is imposed on layer, runs through all the ways that we look at nature. The fundamental particles make nuclei, the nuclei join in atoms, the atoms join in molecules, the molecules join in bases, the bases direct the assembly of amino acids, the amino acids join in proteins. We find again in nature something which seems profoundly to correspond to the way in which our own social relations join us.

The Canyon de Chelly is a kind of microcosm of the cultures, and its high point was reached when the Pueblo people built the great structures just after AD 1000. They represent not only an understanding of nature in the stonework, but of human relations; because the Pueblo people formed here and elsewhere a kind of miniature city. The cliff dwellings were sometimes terraced to five or six storeys, with the top floors recessed from the lower ones. The front of the block was flat with the cliff, the back bowed back into the cliff. These large architectural complexes sometimes have a ground plan of two or three acres, and are made up of four hundred rooms or more.

Stones make a wall, walls make a house, houses make streets, and streets make a city. A city is stones and a city is people; but it is not a heap of stones, and it is not just a jostle of people. In the step from the village to the city, a new community organisation is built, based on the division of labour and on chains of command. The way to recapture that is to walk into the streets of a city that none of us has seen, in a culture that has vanished.

Machu Picchu is in the high Andes, eight thousand feet up in South America. It was built by the Incas at the height of their empire, round about AD 1500 or a little earlier (almost exactly when Columbus reached the West Indies) when the planning of a city was their greatest achievement. When the Spaniards conquered and plundered Peru in 1532, they somehow overlooked Machu Picchu and its sister

The streets of a city that none of us has seen, in a culture that has vanished. *Mortarless joints and cushioned faces of the granite blocks characterise Inca masonry.*

cities. After that it was forgotten for four hundred years, until one winter's day in 1911 Hiram Bingham, a young archaeologist from Yale University, stumbled on it. By then it had been abandoned for centuries and was picked bare as a bone. But in that skeleton of a city lies the structure of every city civilisation, in every age, everywhere in the world.

A city must live on a base, a hinterland, of a rich agricultural surplus; and the visible base for the Inca civilisation was the cultivation of terraces. Of course now the bare terraces grow nothing but grass, but once the potato was cultivated here (it is a native product of Peru), and maize which was long native by then, and in the first place had come from the north. And since this was a ceremonial city of some kind, when the Inca came to visit no doubt there were grown for him tropical luxuries of this climate like the coca, which is an intoxicating herb that only the Inca aristocracy was allowed to chew, and from which we derive cocaine.

At the heart of the terrace culture is a system of irrigation. This is what the pre-Inca empires and Inca empire made; it runs through these terraces, through canals and aqueducts, through the great ravines, down into the desert towards the Pacific and makes it flower. Exactly as in the Fertile Crescent it is the control of water that matters, so here in Peru the Inca civilisation was built on the control of irrigation.

A large system of irrigation extending over an empire requires a strong central authority. It was so in Mesopotamia. It was so in Egypt. It was so in the empire of the Incas. And that means that this city and all the cities here rested on an invisible base of communication by which authority was able to be present and audible everywhere, directing orders from the centre and information towards it. Three inventions sustained the network of authority: the roads, the bridges (in a wild country like this), the messages. They came to a centre here when the Inca was here, and from him they went out of here. They are the three links by which every city is held to every other and which, we suddenly realise, are different in this city.

Roads, bridges, messages in a great empire are always advanced inventions, because if they are cut then authority is cut off and breaks down – in modern times they are typically the first target in a revolution. We know that the Inca gave them much care. Yet on the roads there were no wheels, under the bridges there were no arches, the messages were not in writing. The culture of the Incas had not made these inventions by the year AD 1500. That is because civilisation in America started several thousand years late, and was conquered before it had time to make all the inventions of the Old World.

It seems very strange that an architecture that moved large building stones on rollers could miss the use of the wheel; we forget that what is radical about the wheel is the fixed axle. It seems strange to make suspension bridges and miss the arch. And it seems strangest of all to have a civilisation that kept careful records of numerical information, yet did not put them in writing – the Inca was as illiterate as his poorest citizen, or as the Spanish gangster who overthrew him.

The messages in the form of numerical data came to the Inca on pieces of string called *quipus*. The quipu only records numbers (as knots arranged like our decimal system) and I would dearly like to say, as a mathematician, that numbers are as informative and human a symbolism as words; but they are not. The numbers that described the life of a man in Peru were collected on a kind of punched card in reverse, a braille computer card laid out as a knotted piece of string. When he married, the piece of string was moved to another place in the kinship bundle. Everything that was stored in the Inca's armies, granaries and warehouses was noted on these quipus. The fact is that Peru was already the dreaded metropolis of the future, the memory store in which an empire lists the acts of every citizen, sustains him, assigns him his labours, and puts it all down impersonally as numbers.

It was a remarkably tight social structure. Everyone had a place; everyone was provided for; and everyone – peasant, craftsman or soldier – worked for one man, the supreme Inca. He was the civil head of state and he was also the religious incarnation of godhead. The artisans who lovingly carved a stone to represent the symbol of the link between the sun and its god and king, the Inca, worked for the Inca.

So, necessarily, it was an extraordinarily brittle empire. In less than a hundred years, from 1438 onwards, the Incas had conquered three thousand miles of coastline, almost everything between the Andes and the Pacific. And yet, in 1532 an almost illiterate Spanish adventurer, Francisco Pizarro, rode into Peru with no more than sixty-two terrible horses and a hundred and six foot soldiers; and overnight he conquered the great empire. How? By cutting the top off the pyramid – by capturing the Inca. And from that moment, the empire sagged, and the cities, the beautiful cities, lay bare for the gold plunderer and the vultures.

But, of course, a city is more than a central authority. What is a city? A city is people. A city is alive. It is a community which lives on a base of agriculture, so much richer than in a village, that it can afford to sustain every kind of craftsman and make him a specialist for a lifetime.

The specialists are gone, their work has been destroyed. The men who made Machu Picchu – the goldsmith, the coppersmith, the weaver, the potter – their work has been robbed. The woven fabric has decayed, the bronze has perished, the gold has been stolen. All that remains is the work of the masons, the beautiful craftsmanship of the men who made the city – for the men who make a city are not the Incas but the craftsmen. But naturally, if you work for an Inca (if you work for any one man) his tastes rule you and you make no invention. These men still worked to the end of the empire with the beam; they never invented the arch. Here is a measure of the time lag between the New World and the Old, because this is exactly the point which the Greeks reached two thousand years earlier, and at which they also stopped.

Paestum in Southern Italy was a Greek colony whose temples are older than the Parthenon: they date from about 500 BC. Its river has silted up and it is now separated from the sea by dull salt-flats. But its glory is still spectacular. Although it was ransacked by Saracen pirates in the ninth century, and by Crusaders in the eleventh, Paestum in ruins is one of the marvels of Greek architecture.

Paestum is contemporary with the beginning of Greek mathematics; Pythagoras taught in exile in another Greek colony at Crotone not far from here. Like the mathematics of Peru two thousand years later, the Greek temples were bounded by the straight edge and the set square. The Greeks did not invent the arch either, and therefore their temples are crowded avenues of pillars. They seem open when we see them as ruins, but in fact they are monuments without spaces. That is because they had to be spanned by single beams, and the span that can be sustained by a flat beam is limited by the strength of the beam.

If we picture a beam lying across two columns, then a computer analysis will show the stresses in the beam increase as we move the columns farther apart. The longer the beam, the greater the compression that its weight produces in the top, and the greater the tension it produces in the bottom. And stone is weak in tension; the columns will not fail, because they are compressed, but the beam will fail when the tension becomes too great. It will fail at the bottom unless the columns are kept close together.

The Greeks could be ingenious in making the structure light, for example by using two tiers of columns. But such devices were only makeshifts; in any fundamental sense, the physical limitations of stone could not be overcome without a new invention. Since the Greeks were fascinated by geometry, it is puzzling that they did not conceive the arch. But the fact is that the arch is an engineering invention, and very properly is the discovery of a more practical and plebeian culture than either Greece or Peru.

(*Opposite*) The circle remained the basis of the arch when it went into mass-production in Arab countries.
The Great Mosque at Cordoba.

The aqueduct at Segovia in Spain was built by the Romans about AD 100, in the reign of the emperor Trajan. It carries the waters of the Rio Frio that flows from the high Sierra ten miles away. The aqueduct spans the valley for almost half a mile in more than a hundred double-tiered round arches made of rough-hewn granite blocks, laid without lime or cement. Its colossal proportions so awed the Spanish and Moorish citizens in later and more superstitious ages that they named it El Puente del Diablo, the devil's bridge.

The structure seems to us also prodigious and splendid out of proportion to its function of carrying water. But that is because we get water by turning a tap, and we lightly forget the universal problems of city civilisations. Every advanced culture that concentrates its skilled men in cities depends on the kind of invention and organisation that the Roman aqueduct at Segovia expresses.

The Romans did not invent the arch in the first place in stone, but as a moulded construction made of a kind of concrete. Structurally the arch is simply a method of spanning space which does not load the centre more than the rest; the stress flows outward fairly equally throughout. But for this reason the arch can be made of parts: of separate blocks of stone which the load compresses. In this sense, the arch is the triumph of the intellectual method which takes nature apart and puts the pieces together in new and more powerful combinations.

The Romans always made the arch as a semicircle; they had a mathematical form that worked well, and they were not inclined to experiment. The circle remained the basis of the arch still when it went into mass-production in Arab countries. This is plain in the cloistered, religious architecture that the Moors used; for instance, in the great mosque at Cordoba, also in Spain, built in AD 785 after the Arab conquest. It is a more spacious structure than the Greek temple at Paestum, and yet it has visibly run into similar difficulties; that is, once again it is filled with masonry, which cannot be got rid of without a new invention.

Theoretical discoveries that have radical consequences can usually be seen at once to be striking and original. But practical discoveries, even when they turn out to be far-reaching, often have a look that is more modest and less memorable. A structural innovation to break the limitation of the Roman arch did come, probably from outside Europe, and arrived almost by stealth at first. The invention is a new form of the arch based not on the circle, but on the oval. This does not seem a great change, and yet its effect on the articulation of buildings is spectacular. Of course, a pointed arch is higher, and therefore opens more space and light. But, much more radically, the thrust of the Gothic arch makes it possible to hold the space in a new way, as at Rheims. The load is taken off the walls, which can therefore be pierced with glass, and the total effect is to hang the building like a cage from the arched roof. The inside of the building is open, because the skeleton is outside.

John Ruskin describes the effect of the Gothic arch admirably.

> Egyptian and Greek buildings stand, for the most part, by their own weight and mass, one stone passively incumbent on another; but in the Gothic vaults and traceries there is a stiffness analogous to that of the bones of a limb, or fibres of a tree; an elastic tension and communication of force from part to part, and also a studious expression of this throughout every visible line of the building.

Of all the monuments to human effrontery, there is none to match these towers of tracery and glass that burst into the light of Northern Europe before the year 1200. The construction of these huge, defiant monsters is a stunning achievement of human foresight – or rather, I ought to say, since they were built before any mathematician knew how to compute the forces in them, of human insight. Of course it did not happen without mistakes and some sizeable failures. But what must strike the mathematician most about the Gothic cathedrals is how sound the insight in them was, how smoothly and rationally it progressed from the experience of one structure to the next.

The cathedrals were built by the common consent of townspeople, and for them by common masons. They bear almost no relation to the everyday, useful architecture of the time, and yet in them improvisation becomes invention at every moment. As a matter of mechanics, the design had turned the semicircular Roman arch into the high, pointed Gothic arch in such a way that the stress flows through the arch to the outside of the building. And then in the twelfth century also came the sudden revolutionary turning of that into the half arch: the flying buttress. The stress runs in the buttress as it runs in my arm when I raise my hand and push against the building as if to support it – there is no masonry where there is no stress. No basic principle of architecture was added to that realism until the invention of steel and reinforced concrete buildings.

One has the sense that the men who conceived these high buildings were intoxicated by their new-found command of the force in the stone. How else could they have proposed to build Vaults of 125 feet and 150 feet at a time when they could not calculate any of the stresses? Well, the vault of 150 feet – at Beauvais, less than a hundred miles from Rheims – collapsed. Sooner or later the builders were bound to run into some disaster: there is a physical limit to size, even in cathedrals. And when the roof of Beauvais collapsed in 1284, some years after it was finished, it sobered the high Gothic adventure: no structure as tall as this was attempted again. (Yet the empirical design may have been sound; probably the ground at Beauvais was simply not solid enough, and shifted under the building.) But the vault of 125 feet at Rheims held. And from 1250 onwards Rheims became a centre for the arts of Europe.

The arch, the buttress, the dome (which is a sort of arch in rotation) are not the last steps in bending the grain in nature to our own use. But what lies beyond must have a finer grain: we now have to look for the limits in the material itself. It is as if architecture shifts its focus at the same time as physics does, to the microscopic level of matter. In effect, the modern problem is no longer to design a structure from the materials, but to design the materials for a structure.

The masons carried in their heads a stock, not so much of patterns as of ideas, that grew by experience as they went from one site to the next. They also carried with them a kit of light tools. They marked out with compasses the oval shapes for the vaults and the circles for the rose windows. They defined their intersections with callipers, to line them up and fit them into repeatable patterns. Vertical and horizontal were related by the T-square, as they had been in Greek mathematics, using the right angle. That is, the vertical was fixed with the plumb-line, and the horizontal was fixed, not with a spirit-level, but with a plumb-line joined to a right angle.

The wandering builders were an intellectual aristocracy (like the watchmakers five hundred years later) and could move all over Europe, sure of a job and a welcome; they called themselves freemasons as early as the fourteenth century. The skill that they carried in their hands and their heads seemed to others to be as much a mystery as a tradition, a secret fund of knowledge that stood outside the dreary formalism of pulpit learning that the

The masons carried with them a kit of light tools. The vertical was fixed with the plumb-line; and the horizontal was fixed, not with a spirit level, but with a plumb-line joined to a right angle. *Masons at work, 13th century.*

universities taught. When the work of the freemasons petered out, by the seventeenth century, they began to admit honorary members, who liked to believe that their craft went back to the pyramids. That was not really a flattering legend, because the pyramids were built with a much more primitive geometry than the cathedrals.

Yet there is something in the geometrical vision which is universal. Let me explain my preoccupation with beautiful architectural sites – such as the cathedral at Rheims. What does architecture have to do with science? Particularly, what does it have to do with science the way we used to understand it at the beginning of this century, when science was all numbers – the coefficient of expansion of this metal, the frequency of that oscillator?

The fact of the matter is that our conception of science now, towards the end of the twentieth century, has changed radically. Now we see science as a description and explanation of the underlying structures of nature; and words like structure, pattern, plan, arrangement, architecture constantly occur in every description that we try to make. I have by chance lived with this all my life, and it gives me a special pleasure: the kind of mathematics I have done since childhood is geometrical. However, it is no longer a matter of personal or professional taste, for now that is the everyday language of scientific explanation. We talk about the way crystals are put together, the way atoms are made of their parts – above all we talk about the way that living molecules are made of their parts. The spiral structure of DNA has become the most vivid image of science in the last years. And that imagery lives in these arches.

What did the people do who made this building and others like it? They took a dead heap of stones, which is not a cathedral, and they turned it into a cathedral by exploiting the natural forces of gravity, the way the stone is laid naturally in its bedding planes, the brilliant invention of the flying buttress and arch and so on. And they created a structure that grew out of the analysis of nature into this superb synthesis. The kind of man who is interested in the architecture of nature today is the kind of man who made this architecture nearly eight hundred years ago. There is one gift above all others that makes man unique among the animals, and it is the gift displayed everywhere here: his immense pleasure in exercising and pushing forward his own skill.

A popular cliché in philosophy says that science is pure analysis or reductionism, like taking the rainbow to pieces; and art is pure synthesis, putting the rainbow together. This is not so. All imagination begins by analysing nature. Michelangelo said that vividly, by implication, in his sculpture (it is particularly clear in the sculptures that he did not finish), and he also said it explicitly in his sonnets on the act of creation.

> When that which is divine in us doth try
> To shape a face, both brain and hand unite
> To give, from a mere model frail and slight,
> Life to the stone by Art's free energy.

'Brain and hand unite': the material asserts itself through the hand, and thereby prefigures the shape of the work for the brain. The sculptor, as much as the mason, feels for the form within nature, and for him it is already laid down there. That principle is constant.

> The best of artists hath no thought to show
> Which the rough stone in its superfluous shell
> Doth not include: to break the marble spell
> Is all the hand that serves the brain can do.

By the time Michelangelo carved the head of Brutus, other men quarried the marble for him. But Michelangelo had begun as one of the quarrymen in Carrara, and he still felt that the hammer in their hands and in his was groping in the stone for a shape that was already there.

The quarrymen work in Carrara now for the modern sculptors who come here – Marino Marini, Jacques Lipchitz and Henry Moore. Their descriptions of their work are not as poetic as Michelangelo's, but they carry the same feeling. The reflections of Henry Moore are particularly apposite as they run back to the first genius of Carrara.

> To begin with, as a young sculptor, I could not afford expensive stone, and I got my stone by going round the stone-yards and finding what they would call a 'random block'. Then I had to think in the same way that Michelangelo might have done, so that one had to wait until an idea came that fitted the shape of the stone and that was seen, the idea, in that block.

Of course, it cannot be literally true that what the sculptor imagines and carves out is already there, hidden in the block. And yet the metaphor tells the truth about the relation of discovery that exists between man and nature; and it is characteristic that philosophers of science (Leibniz in particular) have turned to the same metaphor of the mind prompted by a vein in the marble. In one sense, everything that we discover is already there: a sculptured figure and the law of nature are both concealed in the raw material. And in another sense, what a man discovers is discovered *by him*; it would not take exactly the same form in the hands of someone else – neither the sculptured figure nor the law of nature would come out in identical copies when produced by two different minds in two different ages. Discovery is a double relation of analysis and synthesis together. As an analysis, it probes for what is there; but then, as a synthesis, it puts the parts together in a form by which the creative mind transcends the bare limits, the bare skeleton, that nature provides.

Sculpture is a sensuous art. (The Eskimos make small sculptures that are not even meant to be seen, only handled.) So it must seem strange that I choose as my model for science, which is usually thought of as an abstract and cold enterprise, the warm, physical actions of sculpture and architecture. And yet it is right. We have to understand that the world can only be grasped by action, not by contemplation. The hand is more important than the eye. We are not one of those resigned, contemplative civilisations of the Far East or the Middle Ages, that believed that the world has only to be seen and thought about – and who practised no science in the form that is characteristic for us. We are active; and indeed we know, as something more than a symbolic accident in the evolution of man, that it is the hand that drives the subsequent evolution of the brain. We find tools today made by man before he became man. Benjamin Franklin in 1778 called man 'a tool-making animal', and that is right.

I have described the hand when it uses a tool as an instrument of discovery; it is the theme of this essay. We see this every time a child learns to couple hand and tool together – to lace its shoes, to thread a needle, to fly a kite or to play a penny whistle. With the practical action there goes another, namely finding pleasure in the action for its own sake – in the skill that one perfects, and perfects by being pleased with it. This at bottom is responsible for every work of art, and science too: our poetic delight in what human beings do because they can do it. The most exciting thing about that is that the poetic use in the end has the truly profound results. Even in prehistory man already made tools that have an edge finer than they need have. The finer edge in its turn gave the tool a finer use, a practical refinement and extension to processes for which the tool had not been designed.

Henry Moore calls his sculpture *The Knife Edge.* The hand is the cutting edge of the mind. Civilisation is not a collection of finished artefacts, it is the elaboration of processes. In the end, the march of man is the refinement of the hand in action.

The most powerful drive in the ascent of man is his pleasure in his own skill. He loves to do what he does well and, having done it well, he loves to do it better. You see it in his science. You see it in the magnificence with which he carves and builds, the loving care, the gaiety, the effrontery. The monuments are supposed to commemorate kings and religions, heroes, dogmas, but in the end the man they commemorate is the builder.

So the great temple architecture of every civilisation expresses the identification of the individual with the human species. To call it ancestor worship, as in China, is too narrow. The point is that the monument speaks for the dead man to the living, and thereby establishes a sense of permanence which is a characteristically human view: the concept that human life forms a continuity which transcends and flows through the individual. The man buried on his horse or revered in his ship at Sutton Hoo becomes, in the stone monuments of later ages, a spokesman for their belief that there is such an entity as mankind, of which we are each a representative – in life and death.

I could not end this essay without turning to my favourite monuments, built by a man who had no more scientific equipment than a Gothic mason. These are the Watts Towers in Los Angeles, built by an Italian called Simon Rodia. He came from Italy to the United States at the age of twelve. And then at the age of forty-two, having worked as a tile-setter and general repairman, he suddenly decided to build, in his back garden, tremendous structures out of chicken wire, bits of railway tie, steel rods, cement, sea shells, bits of broken glass, and tile of course – anything that he could find or that the neighbourhood children could bring him. It took him thirty-three years to build them. He never had anyone to help him because, he said, 'most of the time I didn't know what to do myself'. He finished them in 1954; he was seventy-five by then. He gave the house, the garden and the towers to a neighbour, and simply walked out.

'I had in mind to do something big,' Simon Rodia had said, 'and I did. You have to be good good or bad bad to be remembered.' He had learned his engineering skill as he went along, by doing, and by taking pleasure in the doing. Of course, the City Building Department decided that the towers were unsafe, and in 1959 they ran a test on them. They tried to pull down one of the towers. I am happy to say that they failed. So the Watts Towers have survived, the work of Simon Rodia's hands, a monument in the twentieth century to take us back to the simple, happy, and fundamental skill from which all our knowledge of the laws of mechanics grows.

The tool that extends the human hand is also an instrument of vision. It reveals the structure of things and makes it possible to put them together in new, imaginative combinations. But, of course, the visible is not the only structure in the world. There is a finer structure below it. And the next step in the ascent of man is to discover a tool to open up the invisible structure of matter.

CHAPTER FOUR

THE HIDDEN STRUCTURE

It is with fire that blacksmiths iron subdue
Unto fair form, the image of their thought:
Nor without fire hath any artist wrought
Gold to its utmost purity of hue.
Nay, nor the unmatched phoenix lives anew,
Unless she burn.

Michelangelo, *Sonnet* 59

What is accomplished by fire is alchemy, whether in the furnace or kitchen stove.

Paracelsus

There is a special mystery and fascination about man's relation to fire, the only one of the four Greek elements that no animal inhabits (not even the salamander). Modern physical science is much concerned with the invisible fine structure of matter, and that is first opened by the sharp instrument of fire. Although that mode of analysis begins several thousand years ago in practical processes (the extraction of salt and of metals, for example) it was surely set going by the air of magic that boils out of the fire: the alchemical feeling that substances can be changed in unpredictable ways. This is the numinous quality that seems to make fire a source of life and a living thing to carry us into a hidden underworld within the material world. Many ancient recipes express it.

> Now the substance of cinnabar is such that the more it is heated, the more exquisite are its sublimations. Cinnabar will become mercury, and passing through a series of other sublimations, it is again turned into cinnabar, and thus it enables man to enjoy eternal life.

This is the classic experiment with which the alchemists in the Middle Ages inspired awe in those who watched them, all the way from China to Spain. They took the red pigment, cinnabar, which is a sulphide of mercury, and heated it. The heat drives off the sulphur and leaves behind an exquisite pearl of the mysterious silvery liquid metal mercury, to astonish and strike awe into the patron. When the mercury is heated in air it is oxidised and becomes, not (as the recipe thought) cinnabar again, but an oxide of mercury that is also red. Yet the recipe was not quite mistaken; the oxide can be turned into mercury again, red to silver, and the mercury to its oxide, silver to red, all by the action of heat.

It is not an experiment of any importance in itself, although it happens that sulphur and mercury are the two elements of which the alchemist before AD 1500 thought the universe was composed. But it does show one important thing, that fire has always been regarded not as the destroying element but as the transforming element. That has been the magic of fire.

I remember Aldous Huxley talking to me through a long evening, his white hands held into the fire, saying, 'This is what transforms. These are the legends that show it. Above all, the legend of the Phoenix that is reborn in the fire, and lives over and over again in generation after generation.' Fire is the image of youth and blood, the symbolic colour in the ruby and cinnabar, and in ochre and haematite with which men painted themselves ceremonially. When Prometheus in Greek mythology brought fire to man, he gave him life and made him into a demigod – that is why the gods punished Prometheus.

In a more practical way, fire has been known to early man for about four hundred thousand years, we think. That implies that fire had already been discovered by *Homo erectus*; as I have stressed, it is certainly found in the caves of Peking man. Every culture since then has used fire, although it is not clear that they all knew how to make fire; in historical times one tribe has been found (the pygmies in the tropical rain forest on the Andaman Islands south of Burma) who carefully tended spontaneous fires because they had no technique for making fire.

In general, the different cultures have used fire for the same purposes: to keep warm, to drive off predators and clear woodland, and to make the simple transformations of everyday life – to cook, to dry and harden wood, to heat and split stones. But, of course, the great transformation that helped to make our civilisation goes deeper: it is the use of fire to disclose a wholly new class of materials, the metals. This is one of the grand technical steps, a stride in the ascent of man, which ranks with the master invention of stone tools; for it was made by discovering in fire a subtler tool for taking matter apart. Physics is the knife that cuts into the grain of nature; fire, the flaming sword, is the knife that cuts below the visible structure, into the stone.

Almost ten thousand years ago, not long after the beginning of the settled communities of agriculture, men in the Middle East began to use copper. But the use of metals could not become general until there was found a systematic process for getting them. That is the extraction of metals from their ores, which we now know was begun rather over seven thousand years ago, about the year 5000 BC in Persia and Afghanistan. At that time, men put the green stone malachite into the fire in earnest, and from it flowed the red metal, copper – happily, copper is released at a modest temperature. They recognised copper because it is sometimes found in raw lumps on the surface, and in that form it had been hammered and worked for over two thousand years already.

The New World too worked copper, and smelted it by the time of Christ, but it paused there. Only the Old World went on to make metal the backbone of civilised life. Suddenly the range of man's control is increased immensely. He has at his command a material which can be moulded, drawn, hammered, cast; which can be made into a tool, an ornament, a vessel; and which can be thrown back into the fire and reshaped. It has only one shortcoming: copper is a soft metal. As soon as it is put under strain, stretched in the form of a wire for instance, it visibly begins to yield. That is because, like every metal, pure copper is made up of layers of crystals. And it is the crystal layers, each like a wafer in which the atoms of the metal are laid out in a regular lattice, which slide over one another until they finally part. When the copper wire begins to neck (that is, develop a weakness), it is not so much that it fails in tension, as that it fails by internal slipping.

Of course the coppersmith did not think like that six thousand years ago. He was faced with a robust problem, which is that copper will not take an edge. For a short time the ascent of man stood poised at the next step: to make a hard metal with a cutting edge. If that seems a large claim for a technical advance, that is because, as a discovery, the next step is so paradoxical and beautiful.

If we picture the next step in modern terms, what needed to be done was plain enough. We have heard that copper as a pure metal is soft because its crystals have parallel planes which easily slip past one another. (It can be hardened somewhat by hammering, to break up the large crystals and make them jagged.) We can deduce that if we could build something gritty into the crystals, that would stop the planes from sliding and would make the metal hard. Of course, on the scale of fine structure that I am describing, something gritty must be a different kind of atoms in place of some of the copper atoms in the crystals. We have to make an alloy whose crystals are more rigid because the atoms in them are not all of the same kind.

That is the modern picture; it is only in the last fifty years that we have come to understand that the special properties of alloys derive from their atomic structure. And yet, by luck or by experiment, the ancient smelters found just this answer: namely, that when to copper you add an even softer metal, tin, you make an alloy which is harder and more durable than either – bronze. Probably the piece of luck was that tin ores in the Old World are found together with copper ores. The point is that almost any pure material is weak, and many impurities will do to make it stronger. What tin does is not a unique but a general function: to add to the pure material a kind of atomic grit – points of a different roughness which stick in the crystal lattices and stop them from sliding.

I have been at pains to describe the nature of bronze in scientific terms because it is a marvellous discovery. And it is marvellous also as a revelation of the potential that a new process carries and evokes in those who handle it. The working of bronze reached its finest expression in China. It had come to China almost certainly from the Middle East, where bronze was discovered about 3800 BC. The high period of bronze in China is also the beginning of Chinese civilisation as we think of it – the Shang dynasty, before 1500 BC.

The Shang dynasty governed a group of feudal domains in the valley of the Yellow River, and for the first time created some unitary state and culture in China. In all ways it is a formative time, when ceramics are also developed and writing becomes fixed. (It is the calligraphy, both on the ceramics and the bronze, which is so startling.) The bronzes in the high period were made with an Oriental attention to detail which is fascinating in itself.

The Chinese made the mould for a bronze casting out of strips shaped round a ceramic core. And because the strips are still found, we know how the process worked. We can follow the preparation of the basic core, the incising of the pattern, and particularly the inscribed lettering on the strips formed on the core. The strips thus make up an outer ceramic mould which is baked hard to take the hot metal. We can even follow the traditional preparation of the bronze. The proportions of copper and tin that the Chinese used are fairly exact. Bronze can be made from almost any proportion between, say, five per cent and twenty per cent of tin added to the copper. But the best Shang bronzes are held at fifteen per cent tin, and there the sharpness of the casting is perfect. At that proportion, bronze is almost three times as hard as copper.

The Shang bronzes are ceremonial, divine objects. They express for China a monumental worship which, in Europe at that same moment, was building Stonehenge. Bronze becomes, from this time onwards, a material for all purposes, the plastic of its age. It has this universal quality wherever it is found, in Europe and in Asia.

But in the climax of the Chinese craftsmanship, the bronze expresses something more. The delight of these Chinese works, vessels for wine and food – in part playful and in part divine – is that they form an art that grows spontaneously out of its own technical skill. The maker is ruled and directed by the material; in shape and in surface, his design flows from the process. The beauty that he creates, the mastery that he communicates, comes from his own devotion to his craft.

The scientific content of these classical techniques is clear-cut. With the discovery that fire will smelt metals comes, in time, the more subtle discovery that fire will fuse them together to make an alloy with new properties. That is as true of iron as of copper. Indeed, the parallel between the metals holds at every stage. Iron also was first used in its natural form; raw iron arrives on the surface of the earth in meteorites, and for that reason its Sumerian name is 'metal from heaven'. When iron ores were smelted later, the metal was recognised because it had already been used. The Indians in North America used meteoric iron, but never could smelt the ores.

Because it is much more difficult to extract from its ores than copper, smelted iron is, of course, a much later discovery. The first positive evidence for its practical use is probably a piece of a tool that has got stuck in one of the pyramids; that gives it a date before 2500 BC. But the wide use of iron was really initiated by the Hittites near the Black Sea around 1500 BC – just the time of the finest bronze in China, the time of Stonehenge.

And as copper comes of age in its alloy, bronze, so iron comes of age in its alloy, steel. Within five hundred years, by 1000 BC, steel is being made in India, and the exquisite properties of different kinds of steel come to be known. Nevertheless, steel remained a special and in some ways a rare material for limited use until quite recent times. As late as two hundred years ago, the steel industry of Sheffield was still small and backward, and the quaker Benjamin Huntsman, wanting to make a precision watch-spring, had to turn metallurgist and discover how to make the steel for it himself.

Since I have turned to the Far East to look at the perfection of bronze, I will take an Oriental example also of the techniques that produce the special properties of steel. They reach their climax, for me, in the making of the Japanese sword, which has been going on in one way or another since AD 800. The making of the sword, like all ancient metallurgy, is surrounded with ritual, and that is for a clear reason. When you have no written language, when you have nothing that can be called a chemical formula, then you must have a precise ceremonial which fixes the sequence of operations so that they are exact and memorable.

So there is a kind of laying on of hands, an apostolic succession, by which one generation blesses and gives to the next the materials, blesses the fire, and blesses the swordmaker. The man who was making this sword holds the title of a 'Living Cultural Monument', formally awarded to the leading masters of ancient arts by the Japanese government. His name is Getsu. And in a formal sense, he is a direct descendant in his craft of the swordmaker Masamune, who perfected the process in the thirteenth century – to repel the Mongols. Or so tradition has it; certainly the Mongols at that time repeatedly tried to invade Japan from China, under the command of the grandson of Genghis Khan, the famous Kublai Khan.

Iron is a later discovery than copper because at every stage it needs more heat – in smelting, working and, naturally, in processing its alloy, steel. (The melting point of iron is about 1500°C, almost 500°C higher than that of copper.) Both in heat treatment and in its response to added elements, steel is a material infinitely more sensitive than bronze. In it, iron is alloyed with a tiny percentage of carbon, less than one per cent usually, and variations in that dictate the underlying properties of the steel.

The process of making the sword reflects the delicate control of carbon and of heat treatment by which a steel object is made to fit its function perfectly. Even the steel billet is not simple, because a sword must combine two different and incompatible properties of materials. It must be flexible, and yet it must be hard. Those are not properties which can be built into the same material unless it consists of layers. In order to achieve that, the steel billet is cut, and then doubled over again and again so as to make a multitude of inner surfaces. The sword that Getsu makes requires him to double the billet fifteen times. This means that the number of layers of steel will be 2^{15}, which is well over thirty thousand layers. Each layer must be bound to the next, which has a different property. It is as if he were trying to combine the flexibility of rubber with the hardness of glass. And the sword, essentially, is an immense sandwich of these two properties.

At the last stage, the sword is prepared by being covered with clay to different thicknesses, so that when it is heated and plunged into water it will cool at different rates. The temperature of the steel for this final moment has to be judged precisely, and in a civilisation in which that is not done by measurement, 'it is the practice to watch the sword being heated until it glows to the colour of the morning sun'. In fairness to the swordmaker, I ought to say that such colour cues were also traditional in steelmaking in Europe: as late as the eighteenth century, the right stage at which to temper steel was when it glowed straw-yellow, or purple, or blue, according to the different use for which it was intended.

The climax, not so much of drama as of chemistry, is the quenching, which hardens the sword and fixes the different properties within it. Different crystal shapes and sizes are produced by the different rates of cooling: large, smooth crystals at the flexible core of the sword, and small jagged crystals at the cutting edge. The two properties of rubber and glass are finally fused in the finished sword. They reveal themselves in its surface appearance – a sheen of shot-silk by which the Japanese set high store. But the test of the sword, the test of a technical practice, the test of a scientific theory, is 'Does it work?' Can it cut the human body in the formal ways that ritual lays down? The traditional cuts are mapped as carefully as the cuts of beef on a diagram in a cookery book: 'Cut number two – the O-jo-dan.' The body is simulated by packed straw, nowadays. But in the past a new sword was tested more literally, by using it to execute a prisoner.

The sword is the weapon of the Samurai. By it they survived endless civil wars that divided Japan from the twelfth century on. Everything about them is fine metalwork: the flexible armour made of steel strips, the horse trappings, the stirrups. And yet the Samurai did not know how to make any of these things themselves. Like the horsemen in other cultures they lived by force, and depended even for their weapons on the skill of villagers whom they alternately protected and robbed. In the long run, the Samurai became a set of paid mercenaries who sold their services for gold.

Our understanding of how the material world is put together from its elements derives from two sources. One, that I have traced, is the development of techniques for making and alloying useful metals. The other is alchemy, and it has a different character. It is small in scale, is not directed to daily uses, and contains a substantial body of speculative theory. For reasons which are oblique but not accidental, alchemy was much occupied with another metal, gold, which is virtually useless. Yet gold has so fascinated human societies that I should be perverse if I did not try to isolate the properties that gave it its symbolic power.

Gold is the universal prize in all countries, in all cultures, in all ages. A representative collection of gold artefacts reads like a chronicle of civilisations. Enamelled gold rosary, 16th century, English. Gold serpent brooch, 400 BC, Greek. Triple gold crown of Abuna, 17th century, Abyssinian. Gold snake bracelet, ancient Roman. Ritual vessels of Achaemenid gold, 6th century BC, Persian. Drinking bowl of Malik gold, 8th century BC, Persian. Bulls' heads in gold ... Ceremonial gold knife, Chimu, Pre-Inca, Peruvian, 9th century ...

Sculpted gold salt-cellar, Benvenuto Cellini, 16th-century figures, made for King Francis I. Cellini recalled what his French patron said of it:

> When I set this work before the king, he gasped in amazement and could not take his eyes off it. He cried in astonishment,

> 'This is a hundred times more heavenly than I would ever have thought! What a marvel the man is!'

The Spaniards plundered Peru for its gold, which the Inca aristocracy had collected as we might collect stamps, with the touch of Midas. Gold for greed, gold for splendour, gold for adornment, gold for reverence, gold for power, sacrificial gold, life-giving gold, gold for tenderness, barbaric gold, voluptuous gold...

The Chinese put their finger on what made it irresistible. Ko Hung said, 'Yellow gold, if melted a hundred times, will not be spoiled.' In that phrase we become aware that gold has a physical quality that makes it singular; which can be tested or assayed in practice, and characterised in theory.

It is easy to see that the man who made a gold artefact was not just a technician, but an artist. But it is equally important, and not so easy to recognise, that the man who assayed gold was also more than a technician. To him gold was an element of science. Having a technique is useful but, like every skill, what brings it to life is its place in a general scheme of nature – a theory.

Men who tested and refined gold made visible a theory of nature: a theory in which gold was unique, and yet might be made from other elements. That is why so much of antiquity spent its time and ingenuity in devising tests for pure gold. Francis Bacon at the opening of the seventeenth century put the issue squarely.

> Gold hath these natures – greatness of weight, closeness of parts, fixation, pliantness or softness, immunity from rust, colour or tincture of yellow. If a man can make a metal that hath all these properties, let men dispute whether it be gold or no.

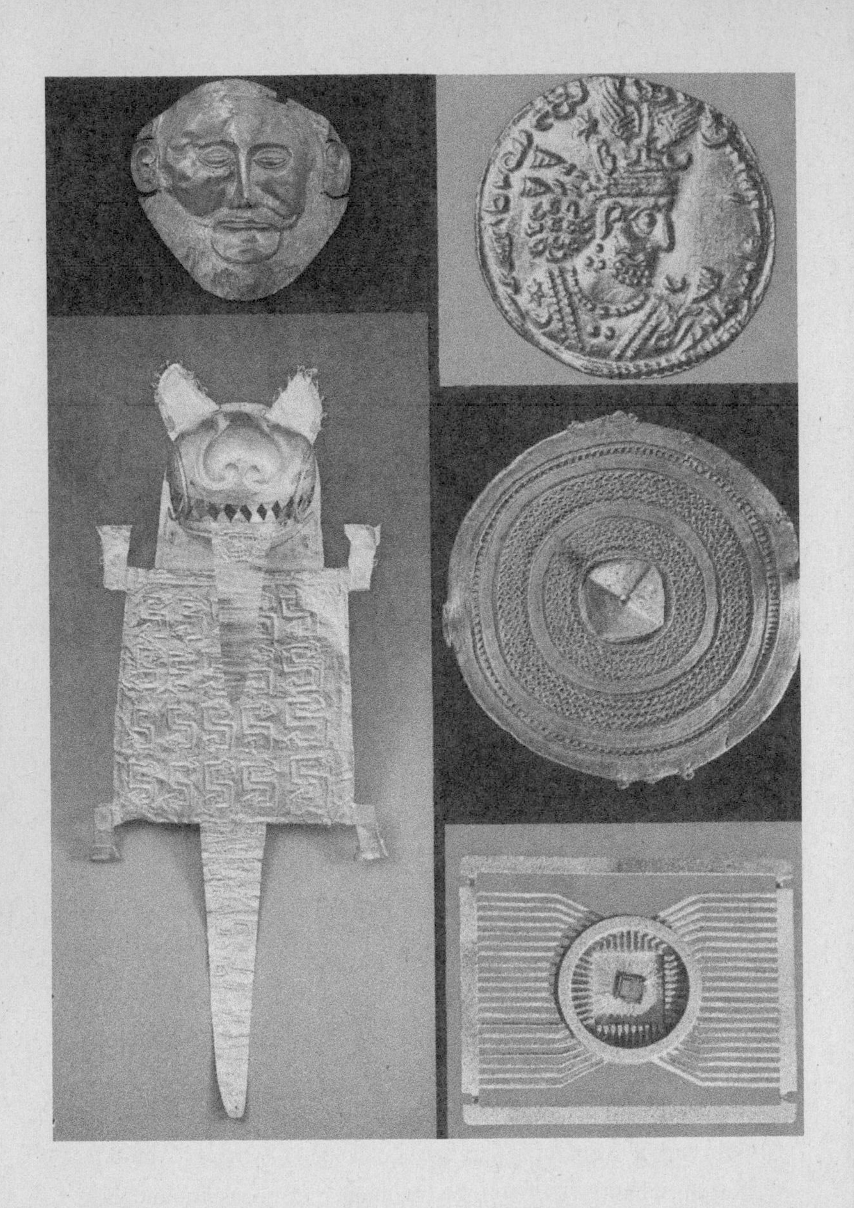

Gold is the universal prize in all countries, in all cultures, in all ages.
Greek gold: Mask of an Achaean king, from a shaft grave in Mycenae, 16th century BC.
Persian gold: Gold dinar of Khusrau II – minted in Iran.
Peruvian gold: Mochica puma, stamped with a design of two-headed serpents.
African gold: Cast gold badge, worn by the Asantehene's (king's) 'soulwasher' as a badge of office, a disk decorated with concentric incised bands, with a pyramidal boss. Ghana, before 1874.
Modern gold: Central input receiver, Concorde Multi-plexing calculator. Edinburgh, 20th century.

Among the several classical tests for gold, one in particular makes the diagnostic property most visible. This is a precise test by cupellation. A bone-ash vessel, or cupel, is heated in the furnace and brought up to a temperature much higher than pure gold requires. The gold, with its impurities or dross, is put in the vessel and melts. (Gold has quite a low melting point, just over 1000°C, almost the same as copper.) What happens now is that the dross leaves the gold and is absorbed into the walls of the vessel: so that all at once there is a visible separation between, as it were, the dross of this world and the hidden purity of the gold in the flame. The dream of the alchemists, to make synthetic gold, has in the end to be tested by the reality of the pearl of gold that survives the assay.

The ability of gold to resist what was called decay (what we would call chemical attack) was singular, and therefore both valuable and diagnostic. It also carried a powerful symbolism, which is explicit even in the earliest formulae. The first written reference we have to alchemy is just over two thousand years old, and comes from China. It tells how to make gold and to use it to prolong life. That is an extraordinary conjunction to us. To us gold is precious because it is scarce; but to the alchemists, all over the world, gold was precious because it was incorruptible. No acid or alkali known to those times would attack it. That indeed is how the emperor's goldsmiths assayed or, as they would have said, parted it, by an acid treatment that was less laborious than cupellation.

When life was thought to be (and for most people was) solitary, poor, nasty, brutish, and short, to the alchemists gold represented the one eternal spark in the human body. Their search to make gold and to find the elixir of life are one and the same endeavour. Gold is the symbol of immortality – but I ought not to say symbol, because in the thought of the alchemists gold was the expression, the embodiment of incorruptibility, in the physical and in the living world together.

So when the alchemists tried to transmute base metals into gold, the transformation that they sought in the fire was from the corruptible to the incorruptible; they were trying to extract the quality of permanence from the everyday. And this was the same as the search for eternal youth: every medicine to fight old age contained gold, metallic gold, as an essential ingredient, and the alchemists urged their patrons to drink from gold cups to prolong life.

Alchemy is much more than a set of mechanical tricks or a vague belief in sympathetic magic. It is from the outset a theory of how the world is related to human life. In a time when there was no clear distinction between substance and process, element and action, the alchemical elements were also aspects of the human personality – just as the Greek elements were also the four humours which the human temperament combines. There lies therefore in their work a profound theory: one which derives in the first place of course from Greek ideas about earth, fire, air and water, but which by the Middle Ages had taken on a new and very important form.

To the alchemists then there was a sympathy between the microcosm of the human body and the macrocosm of nature. A volcano on a grand scale was like a boil; a tempest and rainstorm was like a fit of weeping. Under these superficial analogues lay the deeper concept, which is that the universe and the body are made of the same materials, or principles, or elements. To the alchemists there were two such principles. One was mercury, which stood for everything which is dense and permanent. The other was sulphur, which stood for everything that is inflammable and impermanent. All material bodies, including the human body, were made from these two principles and could be remade from them. For instance, the alchemists believed that all metals grow inside the earth from mercury and sulphur, the way the bones grow inside an embryo from the egg. And they really meant that analogy. It still remains in the symbolism of medicine now. We still use for the female the alchemical sign for copper, that is, what is soft: Venus. And we use for the male the alchemical sign for iron, that is, what is hard: Mars.

That seems a terribly childish theory today, a hodge-podge of fables and false comparisons. But our chemistry will seem childish five hundred years from now. Every theory is based on some analogy, and sooner or later the theory fails because the analogy turns out to be false. A theory in its day helps to solve the problems of the day. And the medical problems had been hamstrung until about 1500, by the belief of the ancients that all cures must come either from plants or from animals – a kind of vitalism which would not entertain the thought that body chemicals are like other chemicals, and which therefore confined medicine largely to herbal cures.

Now the alchemists freely introduced minerals into medicine: salt, for example, was a pivot in the turn-about, and a new theoretician of alchemy made it his third element. He also developed a very characteristic cure for a disease which raged round Europe in 1500 and had not been known before, the new scourge syphilis. To this day we do not know where syphilis came from. It may have been brought back by the sailors in Columbus's ships; it may have spread from the east with the Mongol

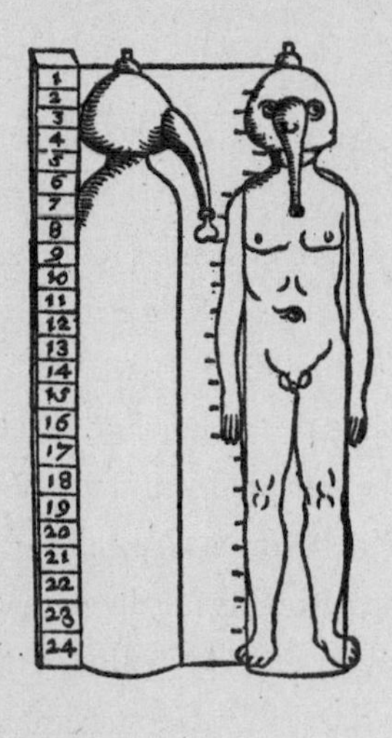

The universe and the body are made of the same materials or principles or elements.

Paracelsus' figure of the furnace of the body with a scale for the study of urine in diagnosis of illness, from the 'Aurora Thesaurusque philosophorum', Basel, 1577.

Paracelsus' figure of the three elements, earth, air and fire.

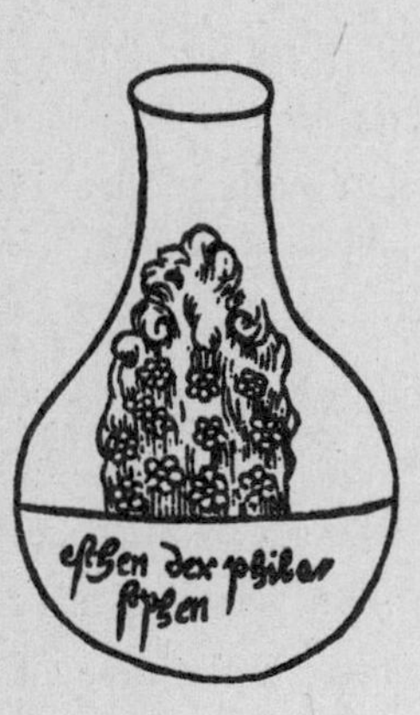

conquests; or it may simply not have been recognised before as a separate disease. The cure for it turned out to depend on the use of the most powerful alchemical metal, mercury. The man who made that cure work is a landmark in the change from the old alchemy to the new, on the way towards modern chemistry: iatrochemistry, biochemistry, the chemistry of life. He worked in Europe in the sixteenth century. The place was Basel in Switzerland. The year was 1527.

There is an instant in the ascent of man when he steps out of the shadowland of secret and anonymous knowledge into a new system of open and personal discovery. The man that I have chosen to symbolise it was christened Aureolus Philippus Theophrastus Bombastus von Hohenheim. Happily, he gave himself the somewhat more compact name of Paracelsus, to publicise his contempt for Celsus and other authors who had been dead more than a thousand years, yet whose medical texts were still current in the Middle Ages. In 1500 the works of classical authors were still thought to contain the inspired wisdom of a golden age, in medicine and science as well as in the arts.

Paracelsus was born near Zürich in 1493, and died at Salzburg in 1541 at the early age of forty-eight. He was a perpetual challenge to everything that was academic: for example, he was the first man to recognise an industrial disease. There are both grotesque and endearing episodes in the undaunted, lifelong battle Paracelsus fought with the oldest tradition of his time, the practice of medicine. His head was a perpetual fountain of theories, many of them contradictory, and most of them outrageous. He was a Rabelaisian, picaresque, wild character, drank with students, ran after women, travelled much over the Old World and, until recently, figured in the histories of science as a quack. But that he was not. He was a man of divided but profound genius.

The point is that Paracelsus was a character. We catch in him, perhaps for the first time, the transparent sense that a scientific discovery flows from a personality, and that discovery comes alive as we watch it being made by a person. Paracelsus was a practical man, who understood that the treatment of a patient depends on diagnosis (he was a brilliant diagnostician) and on direct application by the doctor himself. He broke with the tradition by which the physician was a learned academic who read out of a very old book, and the poor patient was in the hands of some assistant who did what he was told. 'There can be no surgeon who is not also a physician,'

Paracelsus wrote. 'Where the physician is not also a surgeon he is an idol that is nothing but a painted monkey.'

Such aphorisms did not endear Paracelsus to his rivals, but they did make him attractive to other independent minds in the age of the Reformation. That is how he came to be brought to Basel for the single year of triumph in his otherwise disastrous worldly career. In Basel in the year 1527 Johann Frobenius, the great Protestant and humanist printer, had a serious leg infection – the leg was about to be amputated – and in despair appealed to his friends in the new movement, who sent him Paracelsus. Paracelsus threw the academics out of the room, saved the leg, and effected a cure which echoed through Europe. Erasmus wrote to him saying: 'You have brought back Frobenius, who is half my life, from the underworld'.

It is not accidental that new, iconoclastic ideas in medicine and chemical treatment come cheek by jowl, in time and in place, with the Reformation that Luther started in 1517. A focus of that historic time was Basel. Humanism had flourished there even before the Reformation. There was a university with a democratic tradition, so that, although its medical men looked askance at Paracelsus, the City Council could insist that he be allowed to teach. The Frobenius family was printing books, among them some by Erasmus, which spread the new outlook everywhere and in all fields.

A great change was blowing up in Europe, greater perhaps even than the religious and political upheaval that Martin Luther had set going. The symbolic year of destiny was just ahead, 1543. In that year, three books were published that changed the mind of Europe: the anatomical drawings of Andreas Vesalius; the first translation of the Greek mathematics and physics of Archimedes; and the book by Nicolaus Copernicus, *The Revolution of the Heavenly Orbs*, which put the sun at the centre of the heaven and created what is now called the Scientific Revolution.

All that battle between past and future was summarised prophetically in 1527 in a single action outside the Winster at Basel. Paracelsus publicly threw into the traditional student bonfire an ancient medical textbook by Avicenna, an Arab follower of Aristotle.

There is something symbolic about that midsummer bonfire which I will try to conjure into the present. Fire is the alchemist's element by which man is able to cut deeply into the structure of matter. Then is fire itself a form of matter? If you believe that, you have to give it all sorts of impossible properties – such as, that it is lighter than nothing. Two hundred years after Paracelsus, as late as 1730, that is what chemists tried to do in the theory of phlogiston as a last embodiment of material fire. But there is no such substance as phlogiston, just as there is no such principle as the vital principle – because fire is not a material, any more than life is material. Fire is a process of transformation and change, by which material elements are rejoined into new combinations. The nature of chemical processes was only understood when fire itself came to be understood as a process.

That gesture of Paracelsus had said, 'Science cannot look back to the past. There never was a Golden Age.' And from the time of Paracelsus it took another two hundred and fifty years to discover the new element, oxygen, which at last explained the nature of fire, and took chemistry forward out of the Middle Ages. The odd thing is that the man who made the discovery, Joseph Priestley, was not studying the nature of fire, but of another of the Greek elements, the invisible and omnipresent air.

Most of what remains of Joseph Priestley's laboratory is in the Smithsonian Institution in Washington, D.C. And, of course, it has no business to be there. This apparatus ought to be in Birmingham in England, the centre of the Industrial Revolution, where Priestley did his most splendid work. Why is it here? Because a mob drove Priestley out of Birmingham in 1791.

Priestley's story is characteristic of another conflict between originality and tradition. In 1761 he had been invited, at the age of twenty-eight, to teach modern languages at one of the dissenting academies (he was a Unitarian) which took the place of universities for those who were not conformists of the Church of England. Within a year, Priestley was inspired by the lectures in science of one of his fellow teachers to begin a book about electricity; and from that he turned to chemical experiments. He also became excited about the American Revolution (he had been encouraged by Benjamin Franklin) and later the French Revolution. And so, on the second anniversary of the storming of the Bastille, the loyal citizens burned down what Priestley described as one of the most carefully assembled laboratories in the world. He went to America, but was not made welcome. Only his intellectual equals appreciated him; when Thomas Jefferson became President, he told Joseph Priestley, 'Yours is one of the few lives precious to mankind'.

I would like to be able to tell you that the mob that destroyed Priestley's house in Birmingham shattered the dream of a beautiful, lovable, charming man. Alas, I doubt if that would really be true. I do not think Priestley was very lovable, any more than Paracelsus. I suspect that he was a rather difficult, cold, cantankerous, precise, prim, puritanical man. But the ascent of man is not made by lovable people. It is made by people who have two qualities: an immense integrity, and at least a little genius. Priestley had both.

The discovery that he made was that air is not an elementary substance: that it is composed of several gases and that, among those, oxygen – what he called 'dephlogisticated air' – is the one that is essential to the life of animals. Priestley was a good experimenter, and he went forward carefully in several steps. On 1 August 1774 he made some oxygen, and saw to his astonishment how brightly a candle burned in it. In October of that year he went to Paris, where he gave Lavoisier and others news of his finding. But it was not until he himself came back and, on 8 March 1775, put a mouse into oxygen, that he realised how well one breathed in that atmosphere. A day or two after, Priestley wrote a delightful letter in which he said to Franklin: 'Hitherto only two mice and myself have had the privilege of breathing it'.

Priestley also discovered that the green plants breathe out oxygen in sunlight, and so make a basis for the animals who breathe it in. The next hundred years were to show this is crucial; the animals would not have evolved at all if the plants had not made the oxygen first. But in the 1770s nobody had thought about that.

The discovery of oxygen was given meaning by the clear, revolutionary mind of Antoine Lavoisier (who perished in the French Revolution). Lavoisier repeated an experiment of Priestley's which is almost a caricature of one of the classical experiments of alchemy which I described at the beginning of this essay. Both men heated the red oxide of mercury, using a burning glass (the burning glass was fashionable just then), in a vessel in which they could see gas being produced, and could collect it. The gas was oxygen. That was the qualitative experiment; but to Lavoisier it was the instant clue to the idea that chemical decomposition can be quantified.

The idea was simple and radical; run the alchemical experiment in both directions, and measure the quantities that are exchanged exactly. First, in the forward direction: burn mercury (so that it absorbs oxygen) and measure the exact quantity of oxygen that is taken up from a closed vessel between the beginning of the burning and the end. Now turn the process into reverse: take the mercuric oxide that has been made, heat it vigorously and expel the oxygen from it again. Mercury is left behind, oxygen flows into the vessel, and the crucial question is: 'How much?' Exactly the amount that was taken up before. Suddenly the process is revealed for what it is, a material one of coupling and uncoupling fixed quantities of two substances. Essences, principles, phlogiston, have disappeared. Two concrete elements, mercury and oxygen, have really and demonstrably been put together and taken apart.

It might seem a dizzy hope that we can march from the primitive processes of the first coppersmiths and the magical speculations of the alchemists to the most powerful idea in modern science: the idea of the atoms. Yet the route, the firewalker's route, is direct. One step remains beyond the notion of chemical elements that Lavoisier quantified, to its expression in atomic terms by the son of a Cumberland hand-loom weaver, John Dalton.

After the fire, the sulphur, the burning mercury, it was inevitable that the climax of the story should take place in the chill damp of Manchester. Here, between 1803 and 1808, a Quaker schoolmaster called John Dalton turned the vague knowledge of chemical combination, brilliantly illuminated as it had been by Lavoisier, suddenly into the precise modern conception of atomic theory. It was a time of marvellous discovery in chemistry – in those five years ten new elements were found; and yet Dalton was not interested in any of that. He was, to tell the truth, a somewhat colourless man. (He was certainly colour-blind, and the genetic defect of confusing red with green that he described in himself was long called 'Daltonism'.)

Dalton was a man of regular habits, who walked out every Thursday afternoon to play bowls in the countryside. And the things he was interested in were the things of the countryside, the things that still characterise the landscape in Manchester: water, marsh gas, carbon dioxide. Dalton asked himself concrete questions about the way they combine by weight. Why, when water is made of oxygen and hydrogen, do exactly the same amounts always come together to make a given amount of water? Why when carbon dioxide is made, why when methane is made, are there these constancies of weight?

Throughout the summer of 1803 Dalton worked at the question. He wrote: 'An enquiry into the relative weights of the ultimate particles is, as far as I know, entirely new. I have lately been prosecuting this enquiry with remarkable success.' And he thereby realised that the answer must be, Yes, the old-fashioned Greek atomic theory is true. But the atom is not just an abstraction; in a physical sense, it has a weight which characterises this element or that element. The atoms of one element (Dalton called them 'ultimate or elementary particles') are all alike, and are different from the atoms of another element; and one way in which they exhibit the difference between them is physically, as a difference in weight. 'I should apprehend there are a considerable number of what may properly be called elementary particles, which can never be metamorphosed one into another.'

In 1805 Dalton published for the first time his conception of atomic theory, and it went like this. If a minimum quantity of carbon, an atom, combines to make carbon dioxide, it does so invariably with a prescribed quantity of oxygen – two atoms of oxygen.

If water is then constructed from the two atoms of oxygen, each combined with the necessary quantity of hydrogen, it will be one molecule of water from one oxygen atom and one molecule of water from the other.

The weights are right: the weight of oxygen that produces one unit of carbon dioxide will produce two units of water. Now are the weights right for a compound that has no oxygen in it – for marsh gas or methane, in which carbon combines directly with hydrogen? Yes, exactly. If you remove the two oxygen atoms from the single carbon dioxide molecule, and from the two water molecules, then the material balance is precise: you have the right quantities of hydrogen and carbon to make methane.

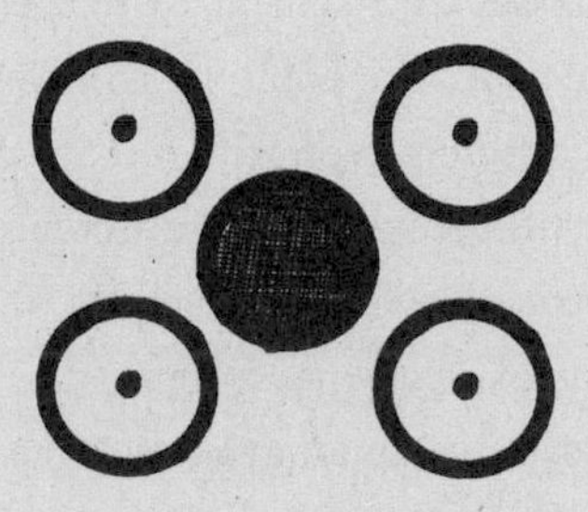

The weighed quantities of different elements that combine with one another express, by their constancy, an underlying scheme of combination between their atoms.

It is the exact arithmetic of the atoms which makes of chemical theory the foundation of modern atomic theory. That is the first profound lesson that comes out of all this multitude of speculation about gold and copper and alchemy, until it reaches its climax in Dalton.

The other lesson makes a point about scientific method. Dalton was a man of regular habits. For fifty-seven years he walked out of Manchester every day; he measured the rainfall, the temperature – a singularly monotonous enterprise in this climate. Of all that mass of data, nothing whatever came. But of the one searching, almost childlike question about the weights that enter the construction of these simple molecules – out of that came modern atomic theory. That is the essence of science: ask an impertinent question, and you are on the way to the pertinent answer.

CHAPTER FIVE

THE MUSIC OF THE SPHERES

Mathematics is in many ways the most elaborated and sophisticated of the sciences – or so it seems to me, as a mathematician. So I find both a special pleasure and constraint in describing the progress of mathematics, because it has been part of so much human speculation: a ladder for mystical as well as rational thought in the intellectual ascent of man. However, there are some concepts that any account of mathematics should include: the logical idea of proof, the empirical idea of exact laws of nature (of space particularly), the emergence of the concept of operations, and the movement in mathematics from a static to a dynamic description of nature. They form the theme of this essay.

Even very primitive peoples have a number system; they may not count much beyond four, but they know that two of any thing plus two of the same thing makes four, not just sometimes but always. From that fundamental step, many cultures have built their own number systems, usually as a written language with similar conventions. The Babylonians, the Mayans, and the people of India, for example, invented essentially the same way of writing large numbers as a sequence of digits that we use, although they lived far apart in space and in time.

So there is no place and no moment in history where I could stand and say 'Arithmetic begins here, now'. People have been counting, as they have been talking, in every culture. Arithmetic, like language, begins in legend. But mathematics in our sense, reasoning with numbers, is another matter. And it is to look for the origin of that, at the hinge of legend and history, that I went sailing to the island of Samos.

In legendary times Samos was a centre of the Greek worship of Hera, the Queen of Heaven, the lawful (and jealous) wife of Zeus. What remains of her temple, the Heraion, dates from the sixth century before Christ. At that time there was born on Samos, about 580 BC, the first genius and the founder of Greek mathematics, Pythagoras. During his lifetime the island was taken over by the tyrant, Polycrates. There is a tradition that before Pythagoras fled, he taught for a while in hiding in a small white cave in the mountains which is still shown to the credulous.

Samos is a magical island. The air is full of sea and trees and music. Other Greek islands will do as a setting for *The Tempest*, but for me this is Prospero's island, the shore where the scholar turned magician. Perhaps Pythagoras was a kind of magician to his followers, because he taught them that nature is commanded by numbers. There is a harmony in nature, he said, a unity in her variety, and it has a language: numbers are the language of nature.

Pythagoras found a basic relation between musical harmony and mathematics. The story of his discovery survives only in garbled form, like a folk tale. But what he discovered was precise. A single stretched string vibrating as a whole produces a ground note. The notes that sound harmonious with it are produced by dividing the string into an exact number of parts: into exactly two parts, into exactly three parts, into exactly four parts, and so on. If the still

Blind harpist, Egypt, 1579-1293 BC

point on the string, the node, does not come at one of these exact points, the sound is discordant.

As we shift the node along the string, we recognise the notes that are harmonious when we reach the prescribed points. Begin with the whole string: this is the ground note. Move the node to the midpoint: this is the octave above it. Move the node to a point one third of the way along: this is the fifth above that. Move it to a point one fourth along: this is the fourth, another octave above. And if you move the node to a point one fifth of the way along, this (which Pythagoras did not reach) is the major third above that.

Pythagoras had found that the chords which sound pleasing to the ear – the western ear – correspond to exact divisions of the string by whole numbers. To the Pythagoreans that discovery had a mystic force. The agreement between nature and number was so cogent that it persuaded them that not only the sounds of nature, but all her characteristic dimensions, must be simple numbers that express harmonies. For example, Pythagoras or his followers believed that we should be able to calculate the orbits of the heavenly bodies (which the Greeks pictured as carried round the earth on crystal spheres) by relating them to the musical intervals. They felt that all the regularities in nature are musical; the movements of the heavens were, for them, the music of the spheres.

These ideas gave Pythagoras the status of a seer in philosophy, almost a religious leader, whose followers formed a secret and perhaps revolutionary sect. It is likely that many of the later followers of Pythagoras were slaves; they believed in the transmigration of souls, which may have been their way of hoping for a happier life after death.

I have been speaking of the language of numbers, that is arithmetic, but my last example was the heavenly spheres, which are geometrical shapes. The transition is not accidental. Nature presents us with shapes: a wave, a crystal, the human body, and it is we who have to sense and find the numerical relations in them. Pythagoras was a pioneer in linking geometry with numbers, and since it is also my choice among the branches of mathematics, it is fitting to watch what he did.

Pythagoras had proved that the world of sound is governed by exact numbers. He went on to prove that the same thing is true of the world of vision. That is an extraordinary achievement. I look about me; here I am, in this marvellous, coloured landscape of Greece, among the wild natural forms, the Orphic dells, the sea. Where under this beautiful chaos can there lie a simple, numerical structure?

The question forces us back to the most primitive constants in our perception of natural laws. To answer well, it is clear that we must begin from universals of experience. There are two experiences on which our visual world is based: that gravity is vertical, and that the horizon stands at right angles to it. And it is that conjunction, those cross-wires in the visual field, which fixes the nature of the right angle; so that if I were to turn this right angle of experience (the direction of 'down' and the direction of sideways') four times, back I come to the cross of gravity and the horizon. The right angle is defined by this fourfold operation, and is distinguished by it from any other arbitrary angle.

In the world of vision, then, in the vertical picture plane that our eyes present to us, a right angle is defined by its fourfold rotation back on itself. The same definition holds also in the horizontal world of experience, in which in fact we move. Consider that world, the world of the flat earth and the map and the points of the compass. Here I am looking across the straits from Samos to Asia Minor, due south. I take a triangular tile as a pointer and I set it pointing there, south. (I have made the pointer in the shape of a right-angled triangle, because I shall want to put its four rotations side by side.) If I turn that triangular tile through a right angle, it points due west. If I now turn it through a second right angle, it points due north. And if I now turn it through a third right angle, it points due east. Finally, the fourth and last turn will take it due south again, pointing to Asia Minor, in the direction in which it began.

Not only the natural world as we experience it, but the world as we construct it is built on that relation. It has been so since the time that the Babylonians built the Hanging Gardens, and earlier, since the time that the Egyptians built the pyramids. These cultures already knew in a practical sense that there is a builder's set square in which the numerical relations dictate and make the right angle. The Babylonians knew many, perhaps hundreds of formulae for this by 2000 BC. The Indians and the Egyptians knew some. The Egyptians, it seems, almost always used a set square with the sides of the triangle made of three, four, and five units. It was not until 550 BC or thereabouts that Pythagoras raised this knowledge out of the world of empirical fact into the world of what we should now call proof. That is, he asked the question, 'How do such numbers that make up these builder's triangles flow from the fact that a right angle is what you turn four times to point the same way?'

His proof, we think, ran something like this. (It is not the proof that stands in the school books.) The four leading points – south, west, north, east – of the triangles that form the cross of the compass are the corners of a square. I slide the four triangles so that the long side of each ends at the leading point of a neighbour. Now I have constructed a square on the longest side of the right-angled triangles – on the hypotenuse. Just so that we should know what is part of the enclosed area and what is not, I will fill in the small inner square area that has now been uncovered with an additional tile. (I use tiles because many tile patterns, in Rome, in the Orient, from now on derive from this kind of wedding of mathematical relation to thought about nature.)

Now we have a square on the hypotenuse, and we can of course relate that by calculation to the squares on the two shorter sides. But that would miss the natural structure and inwardness of the figure. We do not need any calculation. A small game, such as children and mathematicians play, will reveal more than calculation. Transpose two triangles to new positions, thus. Move the triangle that pointed south so that its longest side lies along the longest side of the triangle that pointed north. And move the triangle that pointed east so that its longest side lies along the longest side of the triangle that pointed west.

Now we have constructed an L-shaped figure with the same area (of course, because it is made of the same pieces) whose sides we can see at once in terms of the smaller sides of the rightangled triangle. Let me make the composition of the L-shaped figure visible: put a divider down that separates the end of the L from the upright part. Then it is clear that the end is a square on the shorter side of the triangle; and the upright part of the L is a square on the longer of the two sides enclosing the right angle.

Pythagoras had thus proved a general theorem: not just for the 3 : 4 : 5 triangle of Egypt, or any Babylonian triangle, but for every triangle that contains a right angle. He had proved that the square on the longest side or hypotenuse is equal to the square on one of the other two sides plus the square on the other if, and only if, the angle they contain is a right angle. For instance, the sides 3 : 4 : 5 compose a right-angled triangle because

$$\begin{aligned} 5^2 &= 5 \times 5 = 25 \\ &= 16 + 9 = 4 \times 4 + 3 \times 3 \\ &= 4^2 + 3^2. \end{aligned}$$

And the same is true of the sides of triangles found by the Babylonians, whether simple as 8 : 15 : 17, or forbidding as 3367 : 3456 : 4825, which leave no doubt that they were good at arithmetic.

To this day, the theorem of Pythagoras remains the most important single theorem in the whole of mathematics. That seems a bold and extraordinary thing to say, yet it is not extravagant; because what Pythagoras established is a fundamental characterisation of the space in which we move, and it is the first time that is translated into numbers. And the exact fit of the numbers describes the exact laws that bind the universe. In fact, the numbers that compose right-angled triangles have been proposed as messages which we might send out to planets in other star systems as a test for the existence of rational life there.

The point is that the theorem of Pythagoras in the form in which I have proved it is an elucidation of the symmetry of plane space; the right angle is the element of symmetry that divides the plane four ways. If plane space had a different kind of symmetry, the theorem would not be true; some other relation between the sides of special triangles would be true. And space is just as crucial a part of nature as matter is, even if (like the air) it is invisible; that is what the science of geometry is about. Symmetry is not merely a descriptive nicety; like other thoughts in Pythagoras, it penetrates to the harmony in nature.

When Pythagoras had proved the great theorem, he offered a hundred oxen to the Muses in thanks for the inspiration. It is a gesture of pride and humility together, such as every scientist feels to

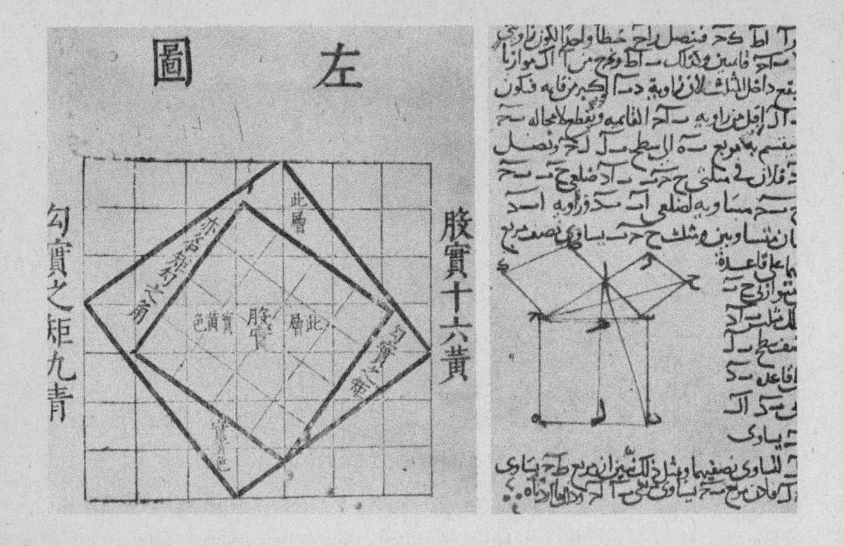

Pythagoras had thus proved a general theorem not just for the 3 : 4 : 5 triangle of Egypt, or any Babylonian triangle, but for every triangle that contains a right angle.

Page from an Arabic version of AD *1258, and a Chinese block print of the theorem.*

this day when the numbers dovetail and say, 'This is a part of, a key to, the structure of nature herself'.

Pythagoras was a philosopher, and something of a religious figure to his followers as well. The fact is there was in him something of that Asiatic influence which flows all through Greek culture and which we commonly overlook. We tend to think of Greece as part of the west; but Samos, the edge of classical Greece, stands one mile from the coast of Asia Minor. From there much of the thought that inspired Greece first flowed; and, unexpectedly, it flowed back to Asia in the centuries after, before ever it reached Western Europe.

Knowledge makes prodigious journeys, and what seems to us a leap in time often turns out to be a long progression from place to place, from one city to another. The caravans carry with their merchandise the methods of trade of their countries – the weights and measures, the methods of reckoning – and techniques and ideas went where they went, through Asia and North Africa. As one example among many, the mathematics of Pythagoras has not come to us directly. It fired the imagination of the Greeks, but the place where it was formed into an orderly system was the Nile city, Alexandria. The man who made the system, and made it famous, was Euclid, who probably took it to Alexandria around 300 BC.

Euclid evidently belonged to the Pythagorean tradition. When a listener asked him what was the practical use of some theorem, Euclid is reported to have said contemptuously to his slave, 'He wants to profit from learning – give him a penny'. The reproof was probably adapted from a motto of the Pythagorean brotherhood, which translates roughly as 'A diagram and a step, not a diagram and a penny' – 'a step' being a step in knowledge or what I have called the ascent of man.

The impact of Euclid as a model of mathematical reasoning was immense and lasting. His book *Elements of Geometry* was translated and copied more than any other book except the Bible right into modern times. I was first taught mathematics by a man who still quoted the theorems of geometry by the numbers that Euclid had given them; and that was not uncommon even fifty years ago, and was the standard mode of reference in the past. When John Aubrey about 1680 wrote an account of how Thomas Hobbes in middle age had suddenly fallen 'in love with geometry' and so with philosophy, he explained that it began when Hobbes happened to see 'in a gentleman's library, Euclid's *Elements* lay open, and 'twas the 47 *Element libri* I'. Proposition 47 in Book 1 of Euclid's *Elements* is the famous theorem of Pythagoras.

The other science practised in Alexandria in the centuries around the birth of Christ was astronomy. Again, we can catch the drift of history in the undertow of legend: when the Bible says that three wise men followed a star to Bethlehem, there sounds in the story the echo of an age when wise men are stargazers. The secret of the heavens that wise men looked for in antiquity was read by a Greek called Claudius Ptolemy, working in Alexandria about AD 150. His work came to Europe in Arabic texts, for the original Greek manuscript editions were largely lost, some in the pillage of the great library of Alexandria by Christian zealots in AD 389, others in the wars and invasions that swept the Eastern Mediterranean throughout the Dark Ages.

The model of the heavens that Ptolemy constructed is wonderfully complex, but it begins from a simple analogy. The moon revolves round the earth, obviously; and it seemed just as obvious to Ptolemy that the sun and the planets do the same. (The ancients thought of the moon and the sun as planets.) The Greeks had believed that the perfect form of motion is a circle, and so Ptolemy made the planets run on circles, or on circles running in their turn on other circles. To us, that scheme of cycles and epicycles seems both simple-minded and artificial. Yet in fact the system was a beautiful and a workable invention, and an article of faith for Arabs and Christians right through the Middle Ages. It lasted for fourteen hundred years, which is a great deal longer than any more recent scientific theory can be expected to survive without radical change.

It is pertinent to reflect here why astronomy was developed so early and so elaborately, and in effect became the archetype for the physical sciences. In themselves, the stars must be quite the most improbable natural objects to rouse human curiosity. The human body ought to have been a much better candidate for early systematic interest. Then why did astronomy advance as a first science ahead of medicine? Why did medicine itself turn to the stars for omens, to predict the favourable and the adverse influences competing for the life of the patient – surely the appeal to astrology is an abdication of medicine as a science? In my view, a major reason is that the observed motions of the stars turned out to be calculable, and from an early time (perhaps 3000 BC in Babylon) lent themselves to mathematics. The pre-eminence of astronomy rests on the peculiarity that it can be treated mathematically; and the progress of physics, and most recently of biology, has hinged equally on finding formulations of their laws that can be displayed as mathematical models.

Every so often, the spread of ideas demands a new impulse. The coming of Islam six hundred years after Christ was the new, powerful impulse. It started as a local event, uncertain in its outcome; but once Mahomet conquered Mecca in AD 630, it took the southern world by storm. In a hundred years, Islam captured Alexandria, established a fabulous city of learning in Baghdad, and thrust its frontier to the east beyond Isfahan in Persia. By AD 730 the Moslem empire reached from Spain and Southern France to the borders of China and India: an empire of spectacular strength and grace, while Europe lapsed in the Dark Ages.

In this proselytising religion, the science of the conquered nations was gathered with a kleptomaniac zest. At the same time, there was a liberation of simple, local skills that had been despised. For instance, the first domed mosques were built with no more sophisticated apparatus than the ancient builder's set square – that is still used. The Masjid-i-Jomi (the Friday Mosque) in Isfahan is one of the statuesque monuments of early Islam. In centres like these, the knowledge of Greece and of the east was treasured, absorbed and diversified.

Mahomet had been firm that Islam was not to be a religion of miracles; it became in intellectual content a pattern of contemplation and analysis. Mohammedan writers depersonalised and formalised the godhead: the mysticism of Islam is not blood and wine, flesh and bread, but an unearthly ecstasy.

> Allah is the light of the heavens and the earth. His light may be compared to a niche that enshrines a lamp, the lamp within a crystal of star-like brilliance, light upon light. In temples which Allah has sanctioned to be built for the remembrance of his name do men praise him morning and evening, men whom neither trade nor profit can divert from remembering him.

One of the Greek inventions that Islam elaborated and spread was the astrolabe. As an observational device, it is primitive; it only measures the elevation of the sun or a star, and that crudely. But by coupling that single observation with one or more star maps, the astrolabe also carried out an elaborate scheme of computations that could determine latitude, sunrise and sunset, the time for prayer and the direction of Mecca for the traveller. And over the star map, the astrolabe was embellished with astrological and religious details, of course, for mystic comfort.

For a long time the astrolabe was the pocket watch and the slide rule of the world. When the poet Geoffrey Chaucer in 1391 wrote a primer to teach his son how to use the astrolabe, he copied it from an Arab astronomer of the eighth century.

Calculation was an endless delight to Moorish scholars. They loved problems, they enjoyed finding ingenious methods to solve them, and sometimes they turned their methods into mechanical devices. A more elaborate ready-reckoner than the astrolabe is the astrological or astronomical computer, something like an automatic calendar, made in the Caliphate of Baghdad in the thirteenth century. The calculations it makes are not deep, an alignment of dials for prognostication, yet it is a testimony to the mechanical skill of those who made it seven hundred years ago, and to their passion for playing with numbers.

The most important single innovation that the eager, inquisitive, and tolerant Arab scholars brought from afar was in writing numbers.

The European notation for numbers then was still the clumsy Roman style, in which the number is put together from its parts by simple addition: for example, 1825 is written as MDCCCXXV, because it is the sum of M=1000, D=500, C+C+C= 100+100+100, XX=10+10, and V=5. Islam replaced that by the modern decimal notation that we still call 'Arabic'. In the note in an Arab manuscript, the numbers in the top row are 18 and 25. We recognise 1 and 2 at once as our own symbols (though the 2 is stood on end). To write 1825, the four symbols would simply be written as they stand, in order, running straight on as a single number; because it is the place in which each symbol stands that announces whether it stands for thousands, or hundreds, or tens, or units.

However, a system that describes magnitude by place must provide for the possibility of empty places. The Arabic notation requires the invention of a zero. The words *zero* and *cipher* are Arab words; so are *algebra*, *almanac*, *zenith*, and a dozen others in mathematics and astronomy. The Arabs brought the decimal system from India about AD 750, but it did not take hold in Europe for another five hundred years after that.

It may be the size of the Moorish Empire that made it a kind of bazaar of knowledge, whose scholars included heretic Nestorian Christians in the east and infidel Jews in the west. It may be a quality in Islam as a religion, which, though it strove to convert people, did not despise their knowledge. In the east the Persian city of Isfahan is its monument. In the west there survives an equally remarkable outpost, the Alhambra in southern Spain.

Seen from the outside, the Alhambra is a square, brutal fortress that does not hint at Arab forms. Inside, it is not a fortress but a palace, and a palace designed deliberately to prefigure on earth the bliss of heaven. The Alhambra is a late construction. It has the lassitude of an empire past its peak, unadventurous and, it thought, safe. The religion of meditation has become sensuous and self-satisfied. It sounds with the music of water, whose sinuous line runs through all Arab melodies, though they are based fair and square on the Pythagorean scale. Each court in turn is the echo and the memory of a dream, through which the Sultan floated (for he did not walk, he was carried). The Alhambra is most nearly the description of Paradise from the Koran.

> Blessed is the reward of those who labour patiently and put their trust in Allah. Those that embrace the true faith and do good works shall be forever lodged in the mansions of Paradise, where rivers will roll at their feet ... and honoured shall they be in the gardens of delight, upon couches face to face. A cup shall be borne round among them from a fountain, limpid, delicious to those who drink ... Their spouses on soft green cushions and on beautiful carpets shall recline.

The Alhambra is the last and most exquisite monument of Arab civilisation in Europe. The last Moorish king reigned here until 1492 when Queen Isabella of Spain was already backing the adventure of Columbus. It is a honeycomb of courts and chambers, and the Sala de las Camas is the most secret place in the palace. Here the girls from the harem came after the bath and reclined, naked. Blind musicians played in the gallery, the eunuchs padded about. And the Sultan watched from above, and sent an apple down to signal to the girl of his choice that she would spend the night with him.

In a western civilisation, this room would be filled with marvellous drawings of the female form, erotic pictures. Not so here. The representation of the human body was forbidden to Mohammedans. Indeed, even the study of anatomy at all was forbidden, and that was a major handicap to Moslem science. So here we find coloured but extraordinarily simple geometric designs. The artist and the mathematician in Arab civilisation have become one. And I mean that quite literally. These patterns represent a high point of the Arab exploration of the subtleties and symmetries of space itself: the flat, two-dimensional space of what we now call the Euclidean plane, which Pythagoras first characterised.

In the wealth of patterns, I begin with a very straightforward one. It repeats a two-leaved motif of dark horizontal leaves, and another of light vertical leaves. The obvious symmetries are translations (that is, parallel shifts of the pattern) and either horizontal or vertical reflections. But note one more delicate point. The Arabs were fond of designs in which the dark and the light units of the pattern are identical. And so, if for a moment you ignore the colours, you can see that you could turn a dark leaf once through a right angle into the position of a neighbouring light leaf. Then, always rotating round the same point of junction, you can turn it into the next position, and (again round the same point) into the next, and finally back on itself. And the rotation spins the whole pattern correctly; every leaf in the pattern arrives at the position of another leaf, however far from the centre of rotation they lie.

Reflection in a horizontal line is a twofold symmetry of the coloured pattern, and so is reflection in a vertical. But if we ignore the colours, we see that there is a fourfold symmetry. It is provided by the operation of rotating through a right angle, repeated four times, by which I earlier proved the theorem of Pythagoras; and thereby the uncoloured pattern becomes in its symmetry like the Pythagorean square.

I turn to a much more subtle pattern. These windswept triangles in four colours display only one very straightforward kind of symmetry, in two directions. You could shift the pattern horizontally or you could shift it vertically into new, identical positions. Being windswept is not irrelevant. It is unusual to find a symmetrical system which does not allow reflection. However, this one does not, because these windswept triangles are all right-handed in movement, and you cannot reflect them without making them left-handed.

Now suppose you neglect the difference between the green, the yellow, the black, and the royal blue, and think of the distinction as simply between dark triangles and light triangles. Then there is also a symmetry of rotation. Fix your attention again on a point of junction: six triangles meet there, and they are alternately dark and light. A dark triangle can be rotated there into the position of the next dark triangle, then into the position of the next, and finally back into the original position – a threefold symmetry which rotates the whole pattern.

And indeed the possible symmetries need not stop there. If you forget about the colours at all, then there is a lesser rotation by which you could move a dark triangle into the space of the light triangle beside it because it is identical in shape. This operation of rotation then goes on into the dark, into the light, into the dark, into the light, and finally back into the original dark triangle – a sixfold symmetry of space which rotates the whole pattern. And

the sixfold symmetry in fact is the one we all know best, because it is a symmetry of the snow crystal.

At this point, the non-mathematician is entitled to ask, 'So what? Is that what mathematics is about? Did Arab professors, do modern mathematicians, spend their time with that kind of elegant game?' To which the unexpected answer is – Well, it is not a game. It brings us face to face with something which is hard to remember, and that is that we live in a special kind of space – three-dimensional, flat – and the properties of that space are unbreakable. In asking what operations will turn a pattern into itself, we are discovering the invisible laws that govern our space. There are only certain kinds of symmetries which our space can support, not only in man-made patterns, but in the regularities which nature herself imposes on her fundamental, atomic structures.

The structures that enshrine, as it were, the natural patterns of space are the crystals. And when you look at one untouched by human hand – say, iceland spar – there is a shock of surprise in realising that it is not self-evident why its faces should be regular. It is not self-evident why they should even be flat planes. This is how crystals come; we are used to their being regular and symmetrical; but why? They were not made that way by man but by nature. That flat face is the way in which the atoms had to come together – and that one, and that one. The flatness, the regularity has been forced by space on matter with the same finality as space gave the Moorish patterns their symmetries that I analysed.

Take a beautiful cube of pyrites. Or to me the most exquisite crystal of all, fluorite, an octahedron. (It is also the natural shape of the diamond crystal.) Their symmetries are imposed on them by the nature of the space we live in – the three dimensions, the flatness within which we live. And no assembly of atoms can break that crucial law of nature. Like the units that compose a pattern, the atoms in a crystal are stacked in all directions. So a crystal, like a pattern, must have a shape that could extend or repeat itself in all directions indefinitely. That is why the faces of a crystal can only have certain shapes; they could not have anything but the symmetries in the patterns. For example, the only rotations that are possible go twice or four times for a full turn, or three times or six times – not more. And not five times. You cannot make an assembly of atoms to make triangles which fit into space regularly five at a time.

Thinking about these forms of pattern, exhausting in practice the possibilities of the symmetries of space (at least in two dimensions), was the great achievement of Arab mathematics. And it has a wonderful finality, a thousand years old. The king, the naked women, the eunuchs and the blind musicians made a marvellous formal pattern in which the exploration of what exists was perfect, but which, alas, was not looking for any change. There is nothing new in mathematics, because there is nothing new in human thought, until the ascent of man moved forward to a different dynamic.

Christianity began to surge back in northern Spain about AD 1000 from footholds like the village of Santillana in a coastal strip which the Moors never conquered. It is a religion of the earth there, expressed in the simple images of the village – the ox, the ass, the Lamb of God. The animal images would be unthinkable in Moslem worship. And not only the animal form is allowed; the Son of God is a child, His mother is a woman and is the object of personal worship. When the Virgin is carried in procession, we are in a different universe of vision: not of abstract patterns, but of abounding and irrepressible life.

When Christianity came to win back Spain, the excitement of the struggle was on the frontier. Here Moors and Christians, and Jews too, mingled and made an extraordinary culture of different faiths. In 1085 the centre of this mixed culture was fixed for a time in the city of Toledo. Toledo was the intellectual port of entry into Christian Europe of all the classics that the Arabs had brought together from Greece, from the Middle East, from Asia.

We think of Italy as the birthplace of the Renaissance. But the conception was in Spain in the twelfth century, and it is symbolised and expressed by the famous school of translators at Toledo, where the ancient texts were turned from Greek (which Europe had forgotten) through Arabic and Hebrew into Latin. In Toledo, amid other intellectual advances, an early set of astronomical tables was drawn up, as an encyclopedia of star positions. It is characteristic of the city and the time that the tables are Christian, but the numerals are Arabic, and are by now recognisably modern.

The most famous of the translators and the most brilliant was Gerard of Cremona, who had come from Italy specifically to find a copy of Ptolemy's book of astronomy, the *Almagest*, and who stayed on in Toledo to translate Archimedes, Hippocrates, Galen, Euclid – the classics of Greek science.

And yet, to me personally, the most remarkable and, in the long run, the most influential man who was translated was not a Greek. That is because I am interested in the perception of objects in space. And that was a subject about which the Greeks were totally wrong. It was understood for the first time about the year AD 1000 by an eccentric mathematician whom we call Alhazen, who was the one really original scientific mind that Arab culture produced. The Greeks had thought that light goes from the eyes to the object. Alhazen first recognised that we see an object because each point of it directs and reflects a ray into the eye. The Greek view could not explain how an object, my hand say, seems to change size when it moves. In Alhazen's account it is clear that the cone of rays that comes from the outline and shape of my hand grows narrower as I move my hand away from you. As I move it towards you, the cone of rays that enters your eye becomes larger and subtends a larger angle. And that, and only that, accounts for the difference in size. It is so simple a notion that it is astonishing that scientists paid almost no attention to it (Roger Bacon is an exception) for six hundred years. But artists attended to it long before that, and in a practical way. The concept of the cone of rays from object to the eye becomes the foundation of perspective. And perspective is the new idea which now revivifies mathematics.

The excitement of perspective passed into art in north Italy, in Florence and Venice, in the fifteenth century. A manuscript of Alhazen's *Optics* in translation in the Vatican Library in Rome is annotated by Lorenzo Ghiberti, who made the famous bronze perspectives for the doors of the Baptistry in Florence. He was not the first pioneer of perspective – that may have been Filippo Brunelleschi – and there were enough of them to form an identifiable school of the Perspectivi. It was a school of thought, for its aim was not simply to make the figures lifelike, but to create the sense of their movement in space.

The movement is evident as soon as we contrast a work by the Perspectivi with an earlier one. Carpaccio's painting of St Ursula leaving a vaguely Venetian port was painted in 1495. The obvious effect is to give to visual space a third dimension, just as the ear about this time hears another depth and dimension in the new harmonies in European music. But the ultimate effect is not so much depth as movement. Like the new music, the picture and its inhabitants are mobile. Above all, we feel that the painter's eye is on the move.

Contrast a fresco of Florence painted a hundred years earlier, about AD 1350. It is a view of the city from outside the walls, and the painter looks naively over the walls and the tops of the houses as if they were arranged in tiers. But this is not a matter of skill; it is a matter of intention. There is no attempt at perspective because the painter thought of himself as recording things, not as they look, but as they are: a God's eye view, a map of eternal truth.

The perspective painter has a different intention. He deliberately makes us step away from any absolute and abstract view. Not so much a place as a moment is fixed for us, and a fleeting moment: a point of view in time more than in space. All this was achieved by exact and mathematical means. The apparatus has been recorded with care by the German artist, Albrecht Dürer, who travelled to Italy in 1506 to learn 'the secret art of perspective'. Dürer of course has himself fixed a moment in time; and if we re-create his scene, we see the artist choosing the dramatic moment. He could have stopped early in his walk round the model. Or he could have moved, and frozen the vision at a later moment. But he chose to open his eye, like a camera shutter, understandably at the strong moment, when he sees the model full face. Perspective is not one point of view; for the painter, it is an active and continuous operation.

In early perspective it was customary to use a sight and a grid to hold the instant of vision. The sighting device comes from astronomy, and the squared paper on which the picture was drawn is now the stand-by of mathematics. All the natural details in which Dürer delights are expressions of the dynamic of time: the ox and the ass, the blush of youth on the cheek of the Virgin. The picture is *The Adoration of the Magi*. The three wise men from the east have found their star, and what it announces is the birth of time.

The chalice at the centre of Dürer's painting was a test-piece in teaching perspective. For example, we have Uccello's analysis of the way the chalice looks; we can turn it on the computer as the perspective artist did. His eye worked like a turntable to follow and explore its shifting shape, the elongation of the circles into ellipses, and to catch the moment of time as a trace in space.

Analysing the changing movement of an object, as I can do on the computer, was quite foreign to Greek and to Islamic minds. They looked always for what was unchanging and static, a timeless world of perfect order. The most perfect shape to them was the circle. Motion must run smoothly and uniformly in circles; that was the harmony of the spheres.

This is why the Ptolemaic system was built up of circles, along which time ran uniformly and imperturbably. But movements in the real world are not uniform. They change direction and speed at every instant, and they cannot be analysed until a mathematics is invented in which time is a variable. That is a theoretical problem in the heavens, but it is practical and immediate on earth – in the flight of a projectile, in the spurting growth of a plant, in the single splash of a drop of liquid that goes through abrupt changes of shape and direction. The Renaissance did not have the technical equipment to stop the picture frame instant by instant. But the Renaissance had the intellectual equipment: the inner eye of the painter, and the logic of the mathematician.

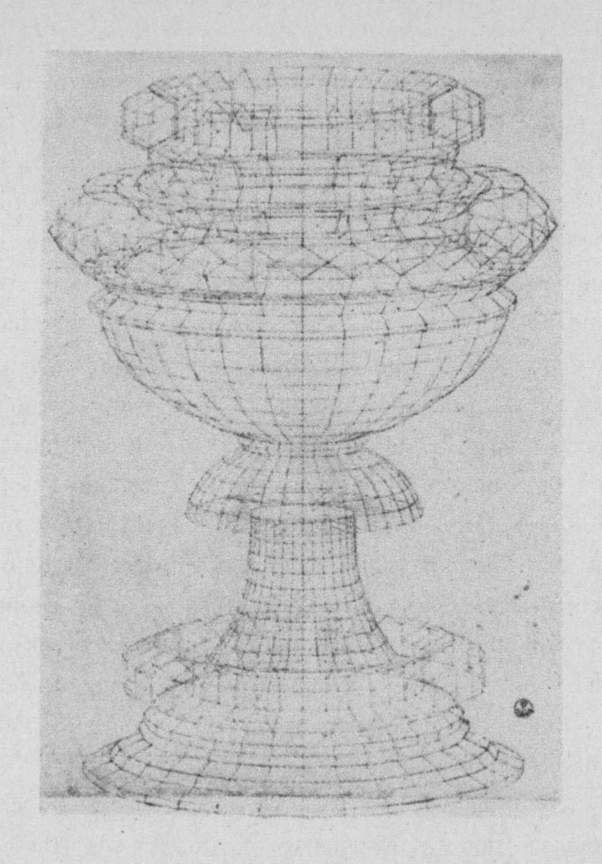

The moment of time as a trace in space.
Paolo Uccello's perspective analysis of a chalice.

In this way Johannes Kepler after the year 1600 became convinced that the motion of a planet is not circular and not uniform. It is an ellipse along which the planet runs at varying speeds. That means that the old mathematics of static patterns will no longer suffice, nor the mathematics of uniform motion. You need a new mathematics to define and operate with instantaneous motion.

The mathematics of instantaneous motion was invented by two superb minds of the late seventeenth century – Isaac Newton and Gottfried Wilhelm Leibniz. It is now so familiar to us that we think of time as a natural element in a description of nature; but that was not always so. It was they who brought in the idea of a tangent, the idea of acceleration, the idea of slope, the idea of infinitesimal, of differential. There is a word that has been forgotten but that is really the best name for that flux of time that Newton stopped like a shutter: *Fluxions* was Newton's name for what is usually called (after Leibniz) the differential calculus. To think of it merely as a more advanced technique is to miss its real content. In it, mathematics becomes a dynamic mode of thought, and that is a major mental step in the ascent of man. The technical concept that makes it work is, oddly enough, the concept of an infinitesimal step; and the intellectual break-through came in giving a rigorous meaning to that. But we may leave the technical concept to the professionals, and be content to call it the mathematics of change.

The laws of nature had always been made of numbers since Pythagoras said that was the language of nature. But now the language of nature had to include numbers which described time. The laws of nature become laws of motion, and nature herself becomes not a series of static frames but a moving process.

CHAPTER SIX

THE STARRY MESSENGER

The first science in the modern sense that grew in the Mediterranean civilisation was astronomy. It is natural to come to astronomy straight from mathematics; after all, astronomy was developed first, and became a model for all the other sciences, just because it could be turned into exact numbers. That is not an idiosyncrasy on my part. What is an idiosyncrasy is that I should choose to begin the drama of the first Mediterranean science in the New World.

The rudiments of astronomy exist in all cultures, and were evidently important in the concerns of early peoples all over the world. One reason for this is clear. Astronomy is the knowledge that guides us through the cycle of the seasons – for example, by the apparent movement of the sun. In this way there can be fixed a time when men should plant, should harvest, move their herds and so on. Therefore all settled cultures have a calendar to guide their plans, and this was true in the New World as it was in the river-basins of Babylon and Egypt.

An example is the civilisation of the Mayans that flourished before AD 1000 in the isthmus of America between the Atlantic and the Pacific Oceans. It has a claim to be the highest of the American cultures: it had a written language, skill in engineering, and original arts. The Mayan temple complexes, with their steep pyramids, housed some astronomers, and we have portraits of a group of them on a large altar stone that has survived. The altar commemorates an ancient astronomical congress that met in the year AD 776. Sixteen mathematicians came here to the famous centre of Mayan science, the sacred city of Copan in Central America.

The Mayans had a system of arithmetic which was far ahead of Europe; for example, they had a symbol for zero. They were good mathematicians; nevertheless, they did not map the motions of the stars, except the simplest. Instead, their ritual was obsessed with the passage of time, and this formal concern dominated their astronomy as it did their poems and legends.

When the great conference met at Copan, the Mayan priest astronomers had run into difficulty. We might suppose that such a major difficulty, calling for learned delegates to come from many centres, would turn on some real problem of observation. But we would be wrong. The congress was called to resolve an arithmetical problem of computation that perpetually troubled the Mayan guardians of the calendar. They kept two calendars, one sacred and one profane, which were never in step for long; and they spent their ingenuity trying to stop the drift between them. The Mayan astronomers had only simple rules for the planetary motions in the heavens, and they had no concept of their machinery. Their idea of astronomy was purely formal, a matter of keeping their calendars right. That was all that was done in AD 776 when the delegates proudly posed for their portraits.

The point is that astronomy does not stop at the calendar. There is another use among early peoples which, however, was not universal. The movements of the stars in the night sky can also serve to guide the traveller, and particularly the traveller at sea who has no other landmarks. That is what astronomy meant to the navigators of the Mediterranean in the Old World. But so far as we can judge, the peoples of the New World did not use astronomy as a scientific guide for land and ocean voyages. And without astronomy it is really not possible to find your way over great distances, or even to have a theory about the shape of the earth and the land and sea on it. Columbus was working with an old and, to our minds, crude astronomy when he set sail for the other side of the world: for instance, he thought that the earth was much smaller than it really is. Yet Columbus found the New World. It cannot be an accident that the New World never thought that the earth is round, and never went out to look for the Old World. It was the Old World which set sail round the earth to discover the New.

Astronomy is not the apex of science or of invention. But it is a test of the cast of temperament and mind that underlies a culture. The seafarers of the Mediterranean since Greek times had a peculiar inquisitiveness that combined adventure with logic – the empirical with the rational – into a single mode of inquiry. The New World did not.

Then did the New World invent nothing? Of course not. Even so primitive a culture as Easter Island made one tremendous invention, the carving of huge and uniform statues. There is nothing like them in the world, and people ask, as usual, all kinds of marginal and faintly irrelevant questions about them. Why were they made like this? How were they transported? How did they get to the places that they are at? But that is not the significant problem. Stonehenge, of a much earlier stone civilisation, was much more difficult to put up; so was Avebury, and many other monuments. No, primitive cultures do inch their way through these enormous communal enterprises.

The critical question about these statues is, Why were they all made *alike*? You see them sitting there, like Diogenes in their barrels, looking at the sky with empty eye-sockets, and watching the sun and the stars go overhead without ever trying to understand them. When the Dutch discovered this island on Easter Sunday in 1722, they said that it had the makings of an earthly paradise. But it did not. An earthly paradise is not made by this empty repetition, like a caged animal going round and round, and making always the same thing. These frozen faces, these frozen frames in a film that is running down, mark a civilisation which failed to take the first step on the ascent of rational knowledge. That is the failure of the New World cultures, dying in their own symbolic Ice Age.

Easter Island is over a thousand miles from the nearest inhabited island, which is Pitcairn Island, to the west. It is over fifteen hundred miles from the next, the Juan Fernandez Islands to the east, where Alexander Selkirk, the original for Robinson Crusoe, was stranded in 1704. Distances like that cannot be navigated unless you have a model of the heavens and of star positions by which you can tell your way. People often ask about Easter Island, How did men come here? They came here by accident: that is not in question. The question is, Why could they not get off? And they could not get off because they did not have a sense of the movement of the stars by which to find their way.

Why not? One obvious reason is that there is no Pole Star in the southern sky. We know that is important, because it plays a part in the migration of birds, which find their way by the Pole Star. That is perhaps why most bird migration is in the northern hemisphere and not in the southern.

The absence of a Pole Star could be meaningful down here in the southern hemisphere, but it cannot be meaningful for the whole of the New World. Because there is Central America, there is Mexico, there are all sorts of places which also did not have an astronomy and yet which lie north of the equator.

What was wrong there? Nobody knows. I think that they lacked that great dynamic image which so moved the Old World – the wheel. The wheel was only a toy in the New World. But in the Old World it was the greatest image of poetry and science; everything was founded on it. This sense of the heavens moving round their hub inspired Christopher Columbus when he set sail in 1492, and the hub was the round earth. He had it from the Greeks, who believed that the stars were fixed on spheres which made music as they turned. Wheels within wheels. That was the system of Ptolemy that had worked for over a thousand years.

More than a hundred years before Christopher Columbus set sail, the Old World had been able to make a superb clockwork of the starry heavens. It was made by Giovanni de Dondi in Padua in about 1350. It took him sixteen years, and it is sad that the original has not survived. Happily, it has been possible to build a duplicate from his working drawings, and the Smithsonian Institution in Washington houses the marvellous model of classical astronomy that Giovanni de Dondi designed.

But more than the mechanical marvel is the intellectual conception, which comes from Aristotle and Ptolemy and the Greeks. De Dondi's clock is their view of the planets as seen from the earth. From the earth there are seven planets – or so the ancients thought, since they counted the sun also as a planet of the earth. So the clock has seven faces or dials, and on each face rides a planet. The path of the planet on its dial is (approximately) the path that we see from the earth – the clock is about as accurate as observation was when it was made. Where the path looks circular from the earth, it is circular on its dial; that was easy. But where the path of a planet loops back on itself when seen from the earth, de Dondi has a mechanical combination of wheels which copies the epicycles (that is, the rolling of circles on circles) by which Ptolemy had described it.

First, then, the Sun: a circular path, as it seemed then. The next dial shows Mars: its motion is running on a clockwork wheel inside a wheel. Then Jupiter: more complex wheels within wheels. Next Saturn: wheels within wheels. Then we come to the Moon – her dial is simple, because she truly is a planet of the earth, and her path is shown as circular. Lastly we come to the dials for the two planets that lie between us and the Sun; that is, to Mercury, and finally to Venus. And again the same picture: the wheel that carries Venus turns inside a larger, hypothetical wheel.

It is a marvellous intellectual conception; very complex – but that only makes it more marvellous that in AD 150, not long after the birth of Christ, the Greeks should have been able to conceive and put into mathematics this superb construction. Then what is wrong with it? One thing only: that there are seven dials for the heavens – and the heavens must have one machinery, not seven. But that machinery was not found until Copernicus put the sun at the centre of the heavens in 1543.

Nicolaus Copernicus was a distinguished churchman and a humanist intellectual from Poland, born in 1473. He had studied law and medicine in Italy; he advised his government on currency reform; and the Pope asked his help on calendar reform. For at least twenty years of his life, roughly, he devoted himself to the modern proposition that nature must be simple. Why were the paths of the planets so complicated? Because, he decided, we look at them from the place where we happen to be standing, the earth. Like the pioneers of perspective, Copernicus asked, Why not look at them from another place? There were good Renaissance reasons, emotional rather than intellectual reasons, that made him choose the golden sun as the other place.

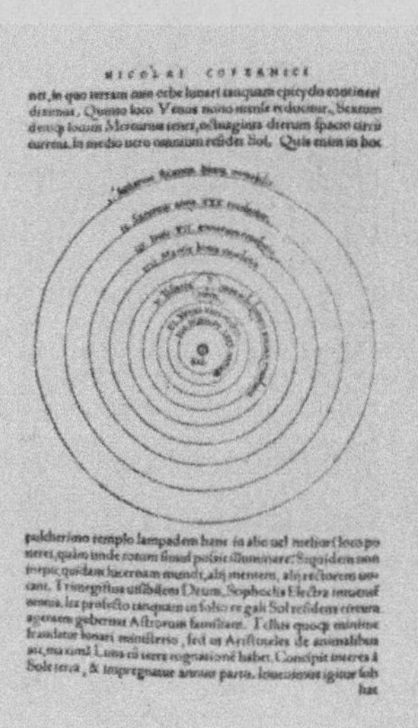

Copernicus put the sun at the centre of the heavens in 1543.
Two pages from the 'De Revolutionibus Orbium Coelestium'.

> In the middle of all sits the Sun enthroned. In this most beautiful temple, could we place this luminary in any better position from which he can illuminate the whole at once? He is rightly called the Lamp, the Mind, the Ruler of the Universe: Hermes Trismegistus names him the Visible God, Sophocles' Electra calls him the All-Seeing. So the Sun sits as upon a royal throne, ruling his children, the planets which circle round him.

We know that Copernicus had thought of putting the sun at the centre of the planetary system for a long time. He may have written the first tentative and non-mathematical sketch of his scheme before he was forty. However, this was not a proposal to be made lightly in an age of religious upheaval. By 1543, near seventy, Copernicus had finally braced himself to publish his mathematical description of the heavens, what he called *De Revolutionibus Orbium Coelestium, The Revolution of the Heavenly Orbs*, as a single system moving round the sun. (The word 'revolution' has an overtone now which is not astronomical, and that is not an accident. It comes from this time and this topic.) Copernicus died in the same year. It is said that he only saw a copy of his book once, when it was put into his hands on his deathbed.

The coming of the Renaissance as a single rush – in religion, art, literature, music, and mathematical science – was a head-on collision with the medieval system as a whole. To us the place of Aristotle's mechanics and Ptolemy's astronomy in the medieval system seems incidental. But to the contemporaries of Copernicus, they represented the natural and visible order of the world. The wheel as the Greek ideal of perfect motion had become a petrified god, as rigid as the Mayan calendar or the figures carved on Easter Island.

The system of Copernicus seemed unnatural to his age, even though the planets still run in circles. (It was a younger man, Johannes Kepler, working later in Prague, who showed that the paths are really elliptical.) That was not what bothered the man in the street, or in the pulpit. They were committed to the wheel of the heavens: the hosts of heaven must march around the earth. That had become an article of faith, as if the Church had made up its mind that the system of Ptolemy was invented not by a Levantine Greek but by the Almighty Himself. Clearly the issue was not one of doctrine but of authority. The issue did not come to a head until seventy years later, in Venice.

Two great men were born in the year 1564; one was William Shakespeare in England, the other was Galileo Galilei in Italy. When Shakespeare writes about the drama of power in his own age, he twice brings the scene to the Republic of Venice: once in *The Merchant of Venice*, and then in *Othello*. That is because in 1600 the Mediterranean was still the centre of the world, and Venice was the hub of the Mediterranean. And here ambitious men came to work, because they were free to work without restraint: merchants, and adventurers, and intellectuals, a host of artists and artisans crowded these streets, as they do now.

The Venetians had the reputation of being a secret and devious people. Venice was a free port, as we would say, and carried with that some of the conspiratorial air which haunts neutral cities like Lisbon and Tangier. It was in Venice that a false patron trapped Giordano Bruno in 1592 and handed him to the Inquisition, which burned him in Rome eight years later.

Certainly the Venetians were a practical people. Galileo had done deep work in fundamental science at Pisa. But what made the Venetians hire him as their professor of mathematics at Padua was, I suspect, his talent for practical inventions. Some of them survive in the historic collection of the *Accademia Cimento* in Florence, and are exquisitely conceived and executed. There is a convoluted glass apparatus for measuring the expansion of liquids, rather like a thermometer; and a delicate hydrostatic balance to find the density of precious objects, on the principle of Archimedes. And there is something which Galileo, who had a knack for salesmanship, called a 'Military Compass', though it is really a calculating instrument not unlike a modern slide-rule. Galileo made and sold them in his own workshop. He wrote a manual for his 'Military Compass' and published it in his own house; it was one of the first works of Galileo to get into print. This was sound, commercial science as the Venetians admired it.

So it is no wonder that when, late in 1608, some spectacle-makers from Flanders invented a primitive form of spyglass, they came to try to sell it to the Republic of Venice. But, of course, the Republic had in its service, in the person of Galileo, a scientist and mathematician immensely more powerful than any in Northern Europe – and a much better publicist who, when he had made a telescope, bustled the Venetian Senate to the top of the Campanile to show it off.

Galileo was a short, square, active man with red hair, and rather more children than a bachelor should have. He was forty-five when he heard the news of the Flemish invention, and it electrified him. He thought it out for himself in one night, and made an instrument about as good, with a magnification of three, which is only about a rather superior opera glass. But before he came to the Campanile in Venice, he stepped the magnification up to eight or ten, and then he had a real telescope. With that, from the top of the Campanile, where the horizon is about twenty miles, you can not only see the ship at sea, you can identify it two hours' sailing and more away. And that was worth a lot of money to the brokers on the Rialto.

Galileo described the events to his brother-in-law in Florence in a letter that he dated 29 August 1609:

> You must know, then, that it is nearly two months since news was spread here that in Flanders there had been presented to Count Maurice a spy-glass, made in such a way that very distant things are made by it to look quite close, so that a man two miles away can be distinctly seen. This seemed to me so marvellous an effect that it gave me occasion for thought; and as it appeared to me that it must be founded on the science of perspective, I undertook to think about its fabrication; which I finally found, and so perfectly that one

which I made far surpassed the reputation of the Flemish one. And word having reached Venice that I had made one, it is six days since I was called by the Signoria, to which I had to show it together with the entire Senate, to the infinite amazement of all; and there have been numerous gentlemen and senators who, though old, have more than once scaled the stairs of the highest campaniles in Venice to observe at sea sails and vessels so far away that, coming under full sail to port, two hours or more were required before they could be seen without my spy-glass. For in fact the effect of this instrument is to represent an object that is, for example, fifty miles away, as large and near as if it were only five.

Galileo is the creator of the modern scientific method. And he did that in the six months following his triumph on the Campanile, which would have been enough for anyone else. It occurred to him then that it was not enough to turn the Flanders toy into an instrument of navigation. It could also be turned into an instrument of research, an idea which was altogether new to that age. He stepped up the magnification of the telescope to thirty, and he turned it on the stars. In that way he really did for the first time what we think of as practical science: build the apparatus, do the experiment, publish the results. And that he did between September of 1609 and March of 1610, when he published in Venice the splendid book *Sidercus Nuncius, The Starry Messenger*, which gave an illustrated account of his new astronomical observations. What did it say?

[I have seen] stars in myriads, which have never been seen before, and which surpass the old, previously known, stars in number more than ten times.

> But that which will excite the greatest astonishment by far, and which indeed especially moved me to call the attention of all astronomers and philosophers, is this, namely, that I have discovered four planets, neither known nor observed by any one of the astronomers before my time.

These were the satellites of Jupiter. *The Starry Messenger* also tells how he turned the telescope on the moon herself. Galileo was the first man to publish maps of the moon. We have his original water-colours.

> It is a most beautiful and delightful sight to behold the body of the moon ... [It] certainly does not possess a smooth and polished surface, but one rough and uneven, and, just like the face of the earth itself, is everywhere full of vast protuberances, deep chasms, and sinuosities.

The British ambassador to the Doge's court in Venice, Sir Henry Wotton, reported to his superiors in England on the day that *The Starry Messenger* came out:

> The mathematical professor at Padua hath ... discovered four new planets rolling about the sphere of Jupiter, besides many other unknown fixed stars; likewise ... that the moon is not spherical, but endued with many prominences ... The author runneth a fortune to be either exceeding famous or exceeding ridiculous. By the next ship your lordship shall receive from me one of the [optical] instruments, as it is bettered by this man.

The news was sensational. It made a reputation larger even than the triumph among the trading community. And yet it was not altogether welcome, because what Galileo saw in the sky, and revealed to everyone who was willing to look, was that the Ptolemaic heaven simply would not work. Copernicus's powerful guess had been right, and now stood open and revealed. And like many more recent scientific results, that did not at all please the prejudice of the establishment of his day.

Galileo thought that all he had to do was to show that Copernicus was right, and everybody would listen. That was his first mistake: the mistake of being naive about people's motives which scientists make all the time. He also thought that his reputation was now large enough for him to be able to go back to his native Florence, leave the rather dreary teaching at Padua which had become burdensome to him, and leave the protection of this essentially anti-clerical, safe Republic of Venice. That was his second and, in the end, fatal mistake.

The successes of the Protestant Reformation in the sixteenth century had caused the Roman Catholic Church to mount a fierce Counter-Reformation. The reaction against Luther was in full cry; the struggle in Europe was for authority. In 1618 the Thirty Years War began. In 1622 Rome created the institution for the propagation of the faith from which we still derive the word *propaganda*. Catholics and Protestants were embattled in what we should now call a cold war, in which, if Galileo had only known it, no quarter was given to a great man or small. The judgment was very simple on both sides: whoever is not for us is – a heretic. Even so unworldly an interpreter of faith as Cardinal Bellarmine had found the astronomical speculations of Giordano Bruno intolerable, and had sent him to the stake. The Church was a great temporal power, and in that bitter time it was fighting a political crusade in which all means were justified by the end – the ethics of the police state.

Galileo seems to me to have been strangely innocent about the world of politics, and most innocent in thinking that he could outwit it because he was clever. For twenty years and more he moved along a path that led inevitably to his condemnation. It took a long time to undermine him; but there was never any doubt that Galileo would be silenced, because the division between him and those in authority was absolute. They believed that faith should dominate; and Galileo believed that truth should persuade.

That clash of principles and, of course, of personalities came into the open at his trial in 1633. But every political trial has a long hidden history of what went on behind the scenes. And the underground history of what came before the trial lies in the locked Secret Archives of the Vatican. Among all these corridors of documents, there is one modest safe in which the Vatican keeps what it regards as the crucial documents. Here, for example, is the application of Henry VIII for divorce – the refusal of which brought the Reformation to England, and ended the tie to Rome. The trial of Giordano Bruno has not left many documents, for the bulk were destroyed; but what exists is here.

And there is the famous Codex 1181, *Proceedings Against Galileo Galilei*. The trial was in 1633. And the first remarkable thing is that the documents begin – when? In 1611, at the moment of Galileo's triumph in Venice, in Florence, and here in Rome, secret information was being laid against Galileo before the Holy Office of the Inquisition. The evidence of the earliest document, not in this file, is that Cardinal Bellarmine instigated inquiries against him. Reports are filed in 1613, 1614, and 1615. By then Galileo himself becomes alarmed. Unbidden, he goes to Rome in order to persuade his friends among the Cardinals not to prohibit the Copernican world system.

But it is too late. In February of 1616, here are the formal words as they stand in draft in the Codex, freely translated:

Propositions to be forbidden:
that the sun is immovable at the centre of the heaven;
that the earth is not at the centre of the heaven, and is not immovable, but moves by a double motion.

Galileo seems to have escaped any severe censure himself. At any rate, he is called before the great Cardinal Bellarmine and he is convinced, and has a letter from Bellarmine to say, that he must not hold or defend the Copernican World System – but there the document stops. Unhappily, there is a document here in the record which goes further, and on which the trial is going to turn. But that is all seventeen years in the future.

Meanwhile Galileo goes back to Florence, and he knows two things. One is that the time to defend Copernicus in public is not yet. And the second, that he thinks that there will be such a time. About the first he is right; about the second, no. However Galileo bided his time, until – when? Until an intellectual Cardinal should be elected Pope: Maffeo Barberini.

That happened in 1623, when Maffeo Barberini became Pope Urban VIII. The new Pope was a lover of the arts. He loved music; he commissioned the composer Gregorio Allegri to write a Miserere for nine voices, which long afterwards was reserved for the Vatican. The new Pope loved architecture. He wanted to make St Peter's the centre of Rome. He put the sculptor and architect Gianlorenzo Bernini in charge of completing the interior of St Peter's, and Bernini boldly designed the tall Baldacchino (the canopy over the Papal throne), which is the only worthy addition to Michelangelo's original design. In his younger days the intellectual Pope had also written poems, one of which was a sonnet of compliments to Galileo on his astronomical writing.

Pope Urban VIII thought of himself as an innovator. He had a confident, impatient turn of mind:

> I know better than all the cardinals put together! The sentence of a living Pope is worth more than all the decrees of a hundred dead ones,

he said imperiously. But in fact, Barberini as Pope turned out to be pure baroque: a lavish nepotist, extravagant, domineering, restless in his schemes, and absolutely tone-deaf to the ideas of others. He even had the birds killed in the Vatican gardens because they disturbed him.

Galileo optimistically came to Rome in 1624, and had six long talks in the gardens with the newly elected Pope. He hoped that the intellectual Pope would withdraw, or at least by-pass, the prohibition of 1616 of the world picture of Copernicus. It turned out that Urban VIII would not consider that. But Galileo still hoped – and the officials of the Papal court expected – that Urban VIII would let the new scientific ideas flow quietly into the Church until, imperceptibly, they replaced the old. After all, that was how the heathen ideas of Ptolemy and Aristotle had become Christian doctrine in the first place. So Galileo went on believing that the Pope was on his side, within the limits set by his office, until it came to the testing time. And then he turned out to be most profoundly mistaken.

Their views had really been intellectually irreconcilable from the beginning. Galileo had always held that the ultimate test of a theory must be found in nature.

> I think that in discussions of physical problems we ought to begin not from the authority of scriptural passages, but from sense-experiences and necessary demonstrations … Nor is God

> any less excellently revealed in Nature's actions than in the sacred statements of the Bible.

Urban VIII objected that there can be no ultimate test of God's design, and insisted that Galileo must say that in his book.

> It would be an extravagant boldness for anyone to go about to limit and confine the Divine power and wisdom to some one particular conjecture of his own.

This proviso was particularly dear to the Pope. In effect, it blocked Galileo from stating any definite conclusion (even the negative conclusion that Ptolemy was wrong), because it would infringe the right of God to run the universe by miracle, rather than by natural law.

The testing time came in 1632 when Galileo finally got his book, the *Dialogue on the Great World Systems*, into print. Urban VIII was outraged.

> Your Galileo has ventured to meddle with things that he ought not to and with the most important and dangerous subjects which can be stirred up in these days,

he wrote to the Tuscan ambassador on 4 September of that year. In the same month came the fateful order:

> His Holiness charges the Inquisitor at Florence to inform Galileo, in the name of the Holy Office, that he is to appear as soon as possible in the course of the month of October at Rome before the Commissary-General of the Holy Office.

The Pope, Maffeo Barberini the friend, Urban VIII, has personally delivered him into the hands of the Holy Office of the Inquisition, whose process is irreversible.

The Dominican cloister of Santa Maria Sopra Minerva was where the Holy Roman and Universal Inquisition proceeded against those whose allegiance was in question. It had been created by Pope Paul III in 1542 to stem the spread of Reformation doctrines, being specially constituted 'against heretical depravity throughout the whole Christian Commonwealth'. After 1571 it had also been given the power to judge written doctrine, and had instituted the Index of Prohibited Books. The rules of procedure were strict and exact. They had been formalised in 1588 and they were, of course, not the rules of a court. The prisoner did not have a copy either of the charges or of the evidence; he had no counsel to defend him.

There were ten judges at the trial of Galileo: all Cardinals and all Dominicans. One of them was the Pope's brother and another was the Pope's nephew. The trial was conducted by the Cornmissar-General of the Inquisition. The hall in which Galileo was tried is now part of the Post Office of Rome, but we know what it looked like in 1633: a ghostly committee room in a club for gentlemen.

We also know exactly the steps by which Galileo came to this pass. It had begun on those walks in the garden with the new Pope in 1624. It was clear that the Pope would not allow the Copernican doctrine to be avowed openly. But there was another way, and the next year Galileo began to write, in Italian, the *Dialogue on the Great World Systems*, in which one speaker put objections to the theory, and the two other speakers, who were rather cleverer, answered them.

Because, of course, the theory of Copernicus is not self-evident. It is not clear how the earth can fly round the sun once a year, or spin on its own axis once a day, and we not fly off. It is not clear how a weight can be dropped from a high tower and fall vertically to a spinning earth. These objections Galileo answered, as it were, on behalf of Copernicus, long dead. We must never forget that Galileo defied the holy establishment in 1616 and in 1633 in defence of a theory not his own, but a dead man's, because he believed it true.

But on his own behalf Galileo put into the book that sense that all his science gives us from the time that, as a young man in Pisa, he had first put his hand on his pulse and watched a pendulum. It is the sense that the laws here on earth reach out into the universe and burst right through the crystal spheres. The forces in the sky are of the same kind as those on earth, that is what Galileo asserts; so that mechanical experiments that we perform here can give us information about the stars. By turning his telescope on the moon, on Jupiter, and on the sunspots, he put an end to the classical belief that the heavens are perfect and unchanging, and only the earth is subject to the laws of change.

The book was finished by 1630, and Galileo did not find it easy to get it licensed. The censors were sympathetic, but it soon became clear that there were powerful forces against the book. However, in the end Galileo collected no fewer than four imprimaturs, and early in 1632 the book was published in Florence. It was an instant success, and for Galileo an instant disaster. Almost at once from Rome the thunder came: Stop the presses. Buy back all the copies – which by then had been sold out. Galileo must come to Rome to answer for it. And nothing that he said could countermand that: his age (he was now nearly seventy), his illness (which was genuine), the patronage of the Grand Duke of Tuscany, nothing counted. He must come to Rome.

It was clear that the Pope himself had taken great umbrage at the book. He had found at least one passage which he had insisted on, put in the book in the mouth of the man who really makes rather the impression of a simpleton. The Preparatory Commission for the trial says so in black and white: that the proviso I have quoted which was so dear to the Pope has been put 'in bocca di un sciocco' – the defender of tradition whom Galileo had named 'Simplicius'. It may be that the Pope felt Simplicius to be a caricature of himself; certainly he felt insulted. He believed that Galileo had hoodwinked him, and that his own censors had let him down.

So, on 12 April 1633, Galileo was brought into this room, sat at this table, and answered the questions from the Inquisitor. The questions were addressed to him courteously in the intellectual atmosphere which reigned in the Inquisition – in Latin, in the third person. How was he brought to Rome? Is this his book? How did he come to write it? What is in his book? All these questions Galileo expected; he expected to defend the book. But then came a question which he did not expect.

Inquisitor: Was he in Rome, particularly in the year 1616, and for what purpose?

Galileo: I was in Rome in the year 1616 because, hearing doubts expressed on the opinions of Nicolaus Copernicus, I came to find out what views it was suitable to hold.

Inquisitor: Let him say what was decided and made known to him then.

Galileo: In the month of February 1616 Cardinal Bellarmine said to me that to hold the opinion of Copernicus as a proven fact was contrary to the Sacred Scriptures. Therefore it could be neither held nor defended; but it could be taken and used as an hypothesis. In confirmation of this I have a certificate from Cardinal Bellarmine, given on 26 May 1616.

Inquisitor: Whether at that time any other precept was given him by someone else?

Galileo: I do not remember anything else that was said or enjoined upon me.

Inquisitor: If it is stated to him that, in the presence of witnesses, there is the instruction that he must not hold or defend the said opinion, or teach it in any way whatsoever, let him now say whether he remembers.

Galileo: I remember that the instruction was that I was neither to hold nor to defend the said opinion. The other two particulars, that is, neither to teach, nor consider in any way whatsoever, they are not stated in the certificate on which I rely.

Inquisitor: After the aforesaid precept, did he obtain permission to write the book?

Galileo: I did not seek permission to write this book because I consider that I did not disobey the instruction I had been given.

Inquisitor: When he asked permission to print the book, did he disclose the command of the Sacred Congregation of which we spoke?

Galileo: I said nothing when I sought permission to publish, not having in the book either held or defended the opinion.

Galileo has a signed document which says that he was forbidden only to hold or defend the theory of Copernicus, which means as if it were a proven matter of fact. That was a prohibition laid on every Catholic at the time. The Inquisition claims that there is a document which prohibits Galileo, and Galileo alone, to teach it *in any way whatsoever* – that is, even by way of discussion or speculation or as a hypothesis. The Inquisition does not have to produce this document. That is not part of the rules of procedure. But we have the document; it is in the Secret Archives, and it is manifestly a forgery – or, at the most charitable, a draft for some suggested meeting which was rejected. It is not signed by Cardinal Bellarmine. It is not signed by the witnesses. It is not signed by the notary. It is not signed by Galileo to show that he received it.

Did the Inquisition really have to stoop to the use of legal quibbles between 'hold or defend', or 'teach in any way whatsoever', in the face of documents which could not have stood up in any court of law? Yes, it did. There was nothing else to do. The book had been published; it had been passed by several censors. The Pope could rage at the censors now – he ruined his own Secretary because he had been helpful to Galileo. But some remarkable public display had to be made to show that the book was to be condemned (it was on the Index for two hundred years) *because of some deceit practised by Galileo*. This was why the trial avoided any matters of substance, either in the book or in Copernicus, and was bent on juggling with formulae and documents. Galileo was to appear deliberately to have tricked the censors, and to have acted not only defiantly but dishonestly.

The court did not meet again; the trial ended here, to our surprise. That is to say, Galileo was twice more brought into this room and allowed to testify on his own behalf; but no questions were asked of him. The verdict was reached at a meeting of the Congregation of the Holy Office over which the Pope presided, which laid down absolutely what was to be done. The dissident scientist was to be humiliated; authority was to be shown large not only in action but in intention. Galileo was to retract; and he was to be shown the instruments of torture as if they were to be used.

What that threat meant to a man who had started life as a doctor we can judge from the testimony of a contemporary who had actually suffered the rack and survived it. That was William Lithgow, an Englishman who had been racked in 1620 by the Spanish Inquisition.

> I was brought to the rack, then mounted on the top of it. My legs were drawn through the two sides of the three-planked rack. A chord was tied about my ankles. As the levers bent forward, the main force of my knees against the two planks burst asunder the sinews of lily, hams, and the lids of my knees were crushed. My eyes began to startle, my mouth to foam and froth, and my teeth to chatter like the doubling of a drummer's sticks. My lips were shivering, lily groans were vehement, and blood sprang from my arms, broken sinews, hands and knees. Being loosed from these pinnacles of pain, I was hand-fast set on the floor, with this incessant imploration: 'Confess! Confess!'

Galileo was not tortured. He was only threatened with torture, twice. His imagination could do the rest. That was the object of the trial, to show men of imagination that they were not immune from the process of primitive, animal fear that was irreversible. But he had already agreed to recant.

> I, Galileo Galilei, son of the late Vincenzo Galilei, Florentine, aged seventy years, arraigned personally before this tribunal, and kneeling before you, most Eminent and Reverend Lord Cardinals, Inquisitors general against heretical depravity throughout the whole Christian Republic, having before my eyes and touching with my hands, the holy Gospels – swear that I have always believed, do now believe, and by God's help will

for the future believe, all that is held, preached, and taught by the Holy Catholic and Apostolic Roman Church. But whereas – after an injunction had been judicially intimated to me by this Holy Office, to the effect that I must altogether abandon the false opinion that the sun is the centre of the world and immovable, and that the earth is not the centre of the world, and moves, and that I must not hold, defend, or teach in any way whatsoever, verbally or in writing, the said doctrine, and after it had been notified to me that the said doctrine was contrary to Holy Scripture – I wrote and printed a book in which I discuss this doctrine already condemned, and adduce arguments of great cogency in its favour, without presenting any solution of these; and for this cause I have been pronounced by the Holy Office to be vehemently suspected of heresy, that is to say, of having held and believed that the sun is the centre of the world and immovable, and that the earth is not the centre and moves.

Therefore, desiring to remove from the minds of your Eminences, and of all faithful Christians, this strong suspicion, reasonably conceived against me, with sincere heart and unfeigned faith I abjure, curse, and detest the aforesaid errors and heresies, and generally every other error and sect whatsoever contrary to the said Holy Church; and I swear that in future I will never again say or assert, verbally or in writing, anything that might furnish occasion for a similar suspicion regarding me; but that should I know any heretic, or person suspected of heresy, I will denounce him to this Holy Office, or to the Inquisitor and ordinary of the place where I may be. Further, I swear and promise to fulfil and observe in their integrity all penances that have been, or that shall be, imposed upon me by this Holy Office. And, in the event of my contravening (which God forbid!) any of these my promises, protestations, and oaths, I submit myself to all the pains and penalties imposed and promulgated in the sacred canons and other constitutions, general and particular, against such delinquents. So help me God, and these His holy Gospels, which I touch with my hands.

I, the said Galileo Galilei, have abjured, sworn, promised, and bound myself as above; and in witness of the truth thereof I have with my own hand subscribed the present document of my abjuration, and recited it word for word at Rome, in the Convent of Minerva, this twenty-second day of June, 1633.

I, Galileo Galilei, have abjured as above with my own hand.

Galileo was confined for the rest of his life in his villa in Arcetri at some distance from Florence, under strict house arrest. The Pope was implacable. Nothing was to be published. The forbidden doctrine was not to be discussed. Galileo was not even to talk to Protestants. The result was silence among Catholic scientists everywhere from then on. Galileo's greatest contemporary, René Descartes, stopped publishing in France and finally went to Sweden.

Galileo made up his mind to do one thing. He was going to write the book that the trial had interrupted: the book on the *New Sciences*, by which he meant physics, not in the stars, but concerning matter here on earth. He finished it in 1636, that is, three years after the trial, an old man of seventy-two. Of course he could not get it published, until finally some Protestants in Leyden in the Netherlands printed it two years later. By that time Galileo was totally blind. He writes of himself:

Alas ... Galileo, your devoted friend and servant, has been for a month totally and incurably blind; so that this heaven, this earth, this universe, which by my remarkable observations and clear demonstrations I have enlarged a hundred, nay, a thousand fold beyond the limits universally accepted by the learned men of all previous ages, are now shrivelled up for me into such a narrow compass as is filled by my own bodily sensations.

Among those who came to see Galileo at Arcetri was the young poet John Milton from England preparing for his life's work, an epic poem that he planned. It is ironic that by the time Milton came to write the great poem, thirty years later, he was totally blind, and he also was dependent on his children to help him finish it.

Milton at the end of his life identified himself with Samson Agonistes, Samson among the Philistines,

> Eyeless in Gaza at the Mill with slaves,

who destroyed the Philistine empire at the moment of his death. And that is what Galileo did, against his own will. The effect of the trial and of the imprisonment was to put a total stop to the scientific tradition in the Mediterranean. From now on the Scientific Revolution moved to Northern Europe. Galileo died, still a prisoner in his house, in 1642. On Christmas Day of the same year, in England, Isaac Newton was born.

CHAPTER SEVEN

THE MAJESTIC CLOCKWORK

When Galileo wrote the opening pages of the *Dialogue on the Great World Systems* about 1630, he said twice that Italian science (and trade) was now in danger of being overtaken by northern rivals. How true a prophecy that was. The man that he had most in mind was the astronomer Johannes Kepler who came to Prague in the year 1600 at the age of twenty-eight and spent his most productive years there. He devised the three laws that turned the system of Copernicus from a general description of the sun and the planets into a precise, mathematical formula.

First, Kepler showed that the orbit of a planet is only roughly circular: it is a broad ellipse in which the sun is slightly off centre, at one focus. Second, a planet does not travel at constant speed: what is constant is the rate at which the line joining the planet to the sun sweeps out the area lying between its orbit and the sun. And third, the time that a particular planet takes for one orbit – its year – increases with its (average) distance from the sun in a quite exact way.

That was the state of affairs when Isaac Newton was born in 1642, that Christmas Day. Kepler had died twelve years earlier, Galileo in that year. And not only astronomy but science stood at a watershed: the coming of a new mind that saw the crucial step from the descriptions that had done duty in the past to the dynamic, causal explanations of the future.

By the year 1650, the centre of gravity of the civilised world had shifted from Italy to Northern Europe. The obvious reason is that the trade routes of the world were different since the discovery and exploitation of America. No longer was the Mediterranean what its name implies, the middle of the world. The middle of the world had shifted north as Galileo had warned, to the fringe of the Atlantic. And with a different trade came a different political outlook, while Italy and the Mediterranean were still ruled by autocracies.

New ideas and new principles now moved forward in the Protestant seafaring nations of the north, England and the Netherlands. England was becoming Republican and Puritan. Dutchmen came over the North Sea to drain the English fens; the marshes became solid land. A spirit of independence grew in the flat vistas and the mists of Lincolnshire, where Oliver Cromwell recruited his Ironsides. By 1650 England was a republic which had cut off the head of its reigning monarch.

When Newton was born at his mother's house in Woolsthorpe in 1642, his father had died some months earlier. In a little while his mother married again, and Newton was left in the care of a grandmother. He was not exactly a homeless boy, and yet from that time he shows none of the intimacy that parents give. All his life he makes the impression of an unloved man. He never married. He never seems to have been able to flow out in that warmth which makes achievement a natural outcome of thought honed in the company of other people. On the contrary, Newton's achievements were solitary, and he always feared that others would steal them from him as (perhaps he thought) they had stolen his mother. We hear almost nothing of him at school or as an undergraduate.

The two years after Newton graduated at Cambridge, 1665 and 1666, were years of Plague, and he spent the times when the University was closed at home. His mother was widowed and back at Woolsthorpe. Here he struck his vein of gold: mathematics. Now that his notebooks have been read, it is clear that Newton had not been well taught, and that he proved most of the mathematics he knew for himself. Then he went on to original discovery. He invented fluxions, what we now call the calculus. Newton kept fluxions as his secret tool; he discovered his results with it, but he wrote them out in conventional mathematics.

Here Newton also conceived the idea of universal gravitation, and at once tested it by calculating the motion of the moon round the earth. The moon was a powerful symbol for him. If she follows her orbit because the earth attracts her, he reasoned, then the moon is like a ball (or an apple) that has been thrown very hard: she is falling towards the earth, but is going so fast that she constantly misses it – she keeps on going round because the earth is round. How great must the force of attraction be?

> I deduced that the forces which keep the planets in their orbs must be reciprocally as the squares of their distances from the centres about which they revolve; and thereby compared the force requisite to keep the moon in her orb with the force of gravity at the surface of the earth; and found them answer pretty nearly.

The understatement is characteristic of Newton; his first rough calculation had, in fact, given the period of the moon close to its true value, about 27¼ days.

When the figures come out right like that, you know as Pythagoras did that a secret of nature is open in the palm of your hand. A universal law governs the majestic clockwork of the heavens, in which the motion of the moon is one harmonious incident. It is a key that you have put into the lock and turned, and nature has yielded in numbers the confirmation of her structure. But, if you are Newton, you do not publish it.

When he went back to Cambridge in 1667, Newton was made a Fellow of his college, Trinity. Two years later his professor resigned the chair of mathematics. It may not have been explicitly in favour of Newton, as used to be thought, but the effect was the same – Newton was appointed. He was then twenty-six.

Newton published his first work in optics. It was conceived like all of his great thought 'in the two plague years of 1665 and 1666, for in those days I was in the prime of my age for invention'. Newton was not at home but had gone back to Trinity College, Cambridge, for a short interval when the Plague slackened.

It is odd to find that a man whom we regard as the master of explanation of the material universe should have begun by thinking about light. There are two reasons for that. First of all, this was a mariner's world, in which the bright minds of England were occupied with all the problems that arose from seafaring. Men like Newton did not think of themselves as doing technical research, of course – that would be too naive an explanation of their interest. They were drawn to the topics that their important elders argued about, as young men have always been. The telescope was a salient problem of the time. And indeed, Newton was first aware of the problem of colours in white light when he was grinding lenses for his own telescope.

But of course, there is beneath this a more fundamental reason. Physical phenomena consist always of the interaction of energy with matter. We see matter by light; we are aware of the presence of light by the interruption by matter. And that thought makes up the world of every great physicist, who finds that he cannot deepen his understanding of one without the other.

In 1666 Newton began to think about what caused the fringes at the edge of a lens, and looked at the effect by simulating it by a prism. Every lens at its edge is a little prism. Now of course the fact that the prism gives you coloured light is a commonplace at least as old as Aristotle. But, alas, so were the explanations of the time, because they made no analysis of quality. They simply said the white light comes through the glass, and it is darkened a little at the thin end, so it only becomes red; it is darkened a little more where the glass is thicker, and becomes green; it is darkened a little more where the glass is thickest, so it becomes blue. Marvellous! For the whole account explains absolutely nothing, yet sounds very plausible. The obvious thing that it does not explain, as Newton pointed out, was self-evident the moment he let the sunlight in through a chink to pass through his prism. It was this: the sun comes in as a circular disc, but it comes out as an elongated shape. Everybody knew that the spectrum was elongated; that also had been known for a thousand years in some way to those who cared to look. But it takes a powerful mind like Newton to break his head on explaining the obvious. And Newton said that the obvious is that the light is not modified; the light is physically separated.

That is a fundamentally new idea in scientific explanation, quite inaccessible to his contemporaries. Robert Hooke argued with him, every kind of physicist argued with him; until Newton got so bored with all the arguments that he wrote to Leibniz,

> I was so persecuted with discussions arising from the publication of my theory of light that I blamed my own imprudence for parting with so substantial a blessing as my quiet to run after a shadow.

From that time on he really refused to have anything to do with debate at all and certainly with the debaters like Hooke. He would not publish his book on optics until 1704, a year after Hooke died, having warned the president of the Royal Society:

> I intend to be no farther solicitous about matters of Philosophy and therefore I hope you will not take it ill if you find me never doing anything more in that kind.

But let us begin at the beginning, in Newton's own words. In the year 1666

> I procured me a Triangular glass-Prisme, to try therewith the celebrated *Phaenomena of Colours*. And in order thereto having darkened my chamber, and made a small hole in my window-shuts, to let in a convenient quantity of the Suns light, I placed my Prisme at his entrance, that it might be thereby refracted to the opposite wall. It was at first a very pleasing divertisement, to view the vivid and intense colours produced thereby; but after a while applying my self to consider them more circumspectly, I became surprised to see them in an *oblong* form; which, according to the received laws of Refraction, I expected should have been *circular*.
>
> And I saw ... that the light, tending to [one] end of the Image, did suffer a Refraction considerably greater then the light tending to the other. And so the true cause of the length of that Image was detected to be no other, then that *Light* consists of *Rays differently refrangible*, which, without any respect to a difference in their incidence, were, according to their degrees of refrangibility, transmitted towards divers parts of the wall.

The elongation of the spectrum was now explained; it was caused by the separation and fanning out of the colours. Blue is bent or refracted more than red, and that is an absolute property of the colours.

> Then I placed another Prisme ... so that the light ... might pass through that also, and be again refracted before it arrived at the wall. This done, I took the first Prisme in my hand and turned it to and fro slowly about its Axis, so much as to make the several parts of the Image ... successively pass through ... that I might observe to what places on the wall the second Prisme would refract them.
>
> When any one sort of Rays hath been well parted from those of other kinds, it hath afterwards obstinately retained its colour, notwithstanding my utmost endeavours to change it.

With that, the traditional view was routed; for if light were modified by glass, the second prism should produce new colours, and turn red to green or blue. Newton called this the critical experiment. It proved that once the colours are separated by refraction, they cannot be changed any further.

> I have refracted it with Prismes, and reflected with it Bodies which in Day-light were of other colours; I have intercepted it with the coloured film of Air interceding two compressed plates of glass; transmitted it through coloured Mediums, and through Mediums irradiated with other sorts of Rays, and diversly terminated it; and yet could never produce any new colour out of it.
>
> But the most surprising, and wonderful composition was that of *Whiteness*. There is no one sort of Rays which alone can exhibit this. 'Tis ever compounded, and to its composition

> are requisite all the aforesaid primary Colours, mixed in a due proportion. I have often with Admiration beheld, that all the Colours of the Prisme being made to converge, and thereby to be again mixed, reproduced light, intirely and perfectly white.
>
> Hence therefore it comes to pass, that *Whiteness* is the usual colour of *Light*; for, Light is a confused aggregate of Rays indued with all sorts of Colors, as they are promiscuously darted from the various parts of luminous bodies.

That letter was written to the Royal Society shortly after Newton was elected a Fellow in 1672. He had shown himself to be a new kind of experimenter, who understood how to form a theory and how to test it decisively against alternatives. He was rather proud of his achievement.

> A naturalist would scarce expect to see ye science of those colours become mathematicall, and yet I dare affirm that there is as much certainty in it as in any other part of Opticks.

Newton had begun to have a reputation in London as well as in the University; and a sense of colour seems to spread into that metropolitan world, as if the spectrum scattered its light across the silks and spices the merchants brought to the capital.

The palette of painters became more varied, there was a taste for richly coloured objects from the East, and it became natural to use many colour words. This is very clear in the poetry of the time. Alexander Pope, who was sixteen when Newton published the *Opticks*, was surely a less sensuous poet than Shakespeare, yet he uses three or four times as many colour words as Shakespeare, and uses them about ten times as often. For instance, Pope's description of fish in the Thames,

> The bright-ey'd Perch with Fins of *Tyrian* Dye,
> The silver Eel, in shining Volumes roll'd,
> The yellow Carp, in Scales bedrop'd with Gold,
> Swift Trouts, diversify'd with Crimson Stains,

would be inexplicable if we did not recognise it as an exercise in colours.

A metropolitan reputation meant, inevitably, new controversies. Results that Newton outlined in letters to London scientists were bandied about. That was how there began, after 1676, a long and bitter dispute with Gonfried Wilhelm Leibniz about priority in the calculus. Newton would never believe that Leibniz, a powerful mathematician himself, had conceived it independently.

Newton thought of retiring altogether from science into his cloister at Trinity. The Great Court was a spacious setting for a scholar in comfortable circumstances; he had his own small laboratory and his own garden. In Neville's Court Wren's great library was being built. Newton subscribed £40 to the fund. It seemed that he might look forward to a donnish life devoted to private study. But, in the end, if he refused to bustle among the scientists in London, they would come to Cambridge to put their arguments to him.

Newton had conceived the idea of a universal gravitation in the Plague year of 1666 and had used it, very successfully, to describe the motion of the moon round the earth. It seems extraordinary that in nearly twenty years that followed he should have made almost no attempt to publish anything about the bigger problem of the motion of the earth round the sun. The stumbling block is uncertain, but the facts are plain. Only in 1684 did there arise in London an argument between Sir Christopher Wren, Robert Hooke and the young astronomer Edmond Halley, as a result of which Halley came to Cambridge to see Newton.

> After they had been some time together, the doctor [Halley] asked him what he thought the curve would be that would be described by the planets, supposing the force of attraction towards the sun to be reciprocal to the square of their distance from it. Sir Isaac replied immediately that it would be an ellipsis. The doctor, struck with joy and amazement, asked him how he knew it. 'Why,' saith he, 'I have calculated it.' Whereupon Dr Halley asked him for his calculation without any further delay. Sir Isaac looked among his papers but could not find it, but he promised him to renew it, and then to send it him.

It took three years, from 1684 to 1687, before Newton wrote out the proof, and it came out as long as – well, in full, as long as the *Principia*. Halley nursed, wheedled, and even financed the *Principia*, and Samuel Pepys accepted it as president of the Royal Society in 1687.

As a system of the world, of course, it was sensational from the moment it was published. It is a marvellous description of the world subsumed under a single set of laws. But much more, it is also a landmark in scientific method. We think of the presentation of science as a series of propositions, one after another, as deriving from the mathematics of Euclid. And so it does. But it is not until Newton turned this into a physical system, by changing mathematics from a static to a dynamic account, that modern scientific method really begins to be rigorous.

And we can see in the book actually where the stumbling blocks were that kept him from pushing on after the orbit of the moon had come out so well. For instance, I am convinced that it is because he could not solve the problem at Section 12 on 'How does a sphere attract a particle?' At Woolsthorpe he had calculated he knew it before you, he ought not blame any but himself, for having taken no more care to secure a discovery, which he puts so much value on. what application he has made in private I know not, but I am sure that the Society have a very great satisfaction in the honour you do them, by your ded[ication] of so worthy a Treatise.

Sr I must now again beg you, not to let your resentments run so high, as to deprive us of your third book, wherein the application of your Mathematicall doctrine to the Theory of Comets, and severall curious Experiments, which, as I guess by what you write, ought to compose it, will undoubtedly render it acceptable to those that will call themselves philosophers without Mathematicks, which are by much the greater number.

Now you approve of the Character and Paper, I will push on the Edition Vigorously. I have sometimes had thoughts of having the Cutts neatly done in Wood, so to stand in the page, with the demonstrations, it will be more convenient, and not much more charge; if it please you to have it so, I will trie how well it can be done; otherwise I will have them in somewhat a larger size than those you have sent up.

I am Sr
Your most affectionate humble Servt
E. Halley

London
June 29. 1686.

It took three years, from 1684 to 1687, before Newton wrote out the proof in full. Halley nursed, wheedled and even financed the *Principia*. *Halley's letter to Isaac Newton when he threatened to abandon the book rather than acknowledge any claim by Robert Hooke, written on 29 June 1686. 'Sir, I must now again beg you not to let your resentments run so high as to deprive us of your third book. Now you approve of the character and paper, I will push on the edition vigorously.'*

roughly, treating the earth and the moon as particles. But they (and the sun and the planets) are large spheres; can the gravitational attraction between them be accurately replaced by an attraction between their centres? Yes, but only (it turned out, ironically) for attractions that fall off as the square of the distance. And in that we see the immense mathematical difficulties that he had to overcome before he could publish.

When Newton was challenged on such questions as 'You have not explained why gravity acts', 'You have not explained how action at a distance could take place', or indeed 'You have not explained why rays of light behave the way they do', he always answered in the same terms: 'I do not make hypotheses'. By which he meant, 'I do not deal in metaphysical speculation. I lay down a law, and derive the phenomena from it'. That was exactly what he had said in his work on optics, and exactly what had not been understood by his contemporaries as a new outlook in optics.

Now if Newton had been a very plain, very dull, very matter-of-fact man, all that would be easily explicable. But I must make you see that he was not. He was really a most extraordinary, wild character. He practised alchemy. In secret, he wrote immense tomes about the Book of Revelation. He was convinced that the law of inverse squares was really already to be found in Pythagoras. And for such a man, who in private was full of these wild metaphysical and mystical speculations, to hold this public face and say, 'I make no hypotheses' – that is an extraordinary expression of his secret character. William Wordsworth in *The Prelude* has a vivid phrase,

> Newton, with his prism and silent face,

which sees and says it exactly.

Well, the public face was very successful of course, Newton could not get promotion in the University, because he was a Unitarian – he did not accept the doctrine of the Trinity, with which scientists in his time were temperamentally ill at ease. Therefore he could not become a parson, therefore he could not possibly become the Master of a College. So, in 1696, Newton went to London to the Mint. In time he became Master of the Mint. After Hooke's death he accepted the Presidency of the Royal Society in 1703. He was knighted by Queen Anne in 1705. And to his death in 1727 he dominated the intellectual landscape of London. The village boy had made good.

The sad thing is that I think he had made good not by his own standards, but only by the standards of the eighteenth century. The sad thing is that it was that society whose criterion he accepted, when he was willing to be a dictator in the councils of the Establishment and count that success.

An intellectual dictator is not a sympathetic figure, even when he has risen from humble beginnings. Yet in his private writings, Newton was not so arrogant as he seems in his public face, so often and so variously represented.

> To explain all nature is too difficult a task for any one man or even for any one age. 'Tis much better to do a little with certainty, and leave the rest for others that come after you, than to explain all things.

And in a more famous sentence he says the same thing less precisely but with a hint of pathos.

> I do not know what I may appear to the world; but to myself I seem to have been only like a boy playing on the sea-shore, and diverting myself in now and then finding a smoother pebble or a prettier shell than ordinary, while the great ocean of truth lay all undiscovered before me.

By the time Newton was in his seventies, little real scientific work was done in the Royal Society. England under the Georges was preoccupied with money (these are the years of the South Sea Bubble), with politics, and with scandal. In the coffee houses, nimble businessmen floated companies to exploit fictitious inventions. Writers poked fun at scientists, in part from spite, and in part from political motives, because Newton was a bigwig in the government establishment.

In the winter of 1713 a group of disgruntled Tory writers formed themselves into a literary society. Until Queen Anne died the next summer, it met often in the rooms in St James's Palace of her physician, Dr John Arbuthnot. The society was called the Scriblerus Club, and set out to ridicule the learned societies of the day. Jonathan Swift's attack on the scientific community in the third book of *Gulliver's Travels* rose out of their discussions. The group of Tories, who later helped John Gay to satirise the government in *The Beggar's Opera*, also helped him in 1717 to write a play *Three Hours After Marriage*. There the butt of the satire is a pompous, ageing scientist under the name of Dr Fossile. Here are some typical scenes from the play between him and an adventurer, Plotwell, who is having an affair with the lady of the house.

Fossile: I promis'd Lady Longfort my eagle-stone. The poor lady is like to miscarry, and 'tis well I thought on't. Hah! Who is here! I do not like the aspect of the fellow. But I will not be over censorious.

Plotwell: Illustrissime domine, hue adveni —

Fossile: Illustrissime domine – non usus sum loquere Latinam – If you cannot speak English, we can have no lingual conversation.

Plotwell: I can speak but a little Englise. I have great deal heard of de fame of de great luminary of all arts and sciences, de illustrious doctor Fossile. I would make commutation (what do you call it), I would exchange some of my tings for some of his tings.

The first topic of fun, naturally, is alchemy; the technical jargon is quite correct throughout.

Fossile: Pray, Sir, what university are you of?
Plotwell: De famous university of Cracow...
Fossile: ... But what Arcanaare you master of, Sir?
Plotwell: See dere, Sir, dat box de snuffi
Fossile: Snuff-box.
Plotwell: Right. Snuff-box. Dat be de very true gold.
Fossile: What of that?
Plotwell: Vat of dat? Me make dat gold my own self, of de lead of de great church of Cracow.
Fossile: By what operations?
Plotwell: By calcination; reverberation; purification; sublimation; amalgumation; precipitation; volitilization.
Fossile: Have a care what you assert. The volitilization of gold is not an obvious process...
Plotwell: I need not acquaint de illustrious doctor Fossile, dat all de metals be but unripe gold.
Fossile: Spoken like a philosopher. And therefore there should be an act of parliament against digging of lead mines, as against felling young timber.

The scientific references come quick and fast now: to the troublesome problem of finding the longitude at sea, to the invention of fluxions or the differential calculus,

Fossile:	I am not at present dispos'd for experiments.
Plotwell:	... Do you deal in longitudes, Sir?
Fossile:	I deal not in impossibilities. I search only for the grand elixir.
Plotwell:	Vat do you tink of de new metode of fluxion?
Fossile:	I know no other but by mercury.
Plotwell:	Ha, ha. Me mean de fluxion of de quantity.
Fossile:	The greatest quantity I ever knew was three quarts a day.
Plotwell:	Be dere any secret in the hydrology, zoology, minerology, hydraulicks, acausticks, pneumaticks, logarithmatechny, dat you do want de explanation of?
Fossile:	This is all out of my way.

It seems irreverent to us that Newton should have been subject to satire in his lifetime, and subject to serious criticism too. But the fact is that every theory, however majestic, has hidden assumptions which are open to challenge and, indeed, in time will make it necessary to replace it. And Newton's theory, beautiful as an approximation to nature, was bound to have the same defect. Newton confessed it. The prime assumption he made is this: that he said at the outset, 'I take space to be absolute'. By that he meant that space is everywhere flat and infinite as it is in our own neighbourhood. And Leibniz criticised that from the outset, and rightly. After all, it is not even probable in our own experience. We are used to living locally in a flat space, but as soon as we look in the large at the earth, we know it not to be so overall.

The earth is spherical; so that the point at the North Pole can be sighted by two observers on the equator who are far apart, yet each of whom says, 'I am looking due North'. Such a state of affairs is inconceivable to an inhabitant of a flat earth, or one who believes that the earth is as flat overall as it seems to be near him. Newton was really behaving like a flat-earther on a cosmic scale: sailing out into space with his foot-rule in one hand and his pocket-watch in the other, mapping space as if it were everywhere as it is here. And that is not necessarily so.

It is not even as if space has to be spherical everywhere – that is, must have a positive curvature. It might well be that space is locally lumpy and undulating. We can conceive of a kind of space that has saddle-points in it, over which massive bodies slide in some directions more easily than in others. The motions of the heavenly bodies must still be the same, of course – we see them and our explanations must fit them. But the explanations would then be different in kind. The laws that govern the moon and the planets would be geometrical and not gravitational.

At that time they were all speculations far in the future, and even if they had been uttered, the mathematics of the day could not cope with them. But thoughtful and philosophic minds were aware that, in laying out space as an absolute grid, Newton had given an unreal simplicity to our perception of things. In contrast, Leibniz had said the prophetic words, 'I hold space to be something purely relative, as time is'.

Time is the other absolute in Newton's system. Time is crucial to mapping the heavens: we do not know in the first place how far away the stars are, only at what moment they pass across our line of sight. So the mariner's world called for the perfection of two sets of instruments: telescopes and clocks.

First, then, improvements in the telescope. They were now centred in the new Royal Observatory at Greenwich. The ubiquitous Robert Hooke had planned that when he was rebuilding London with Sir

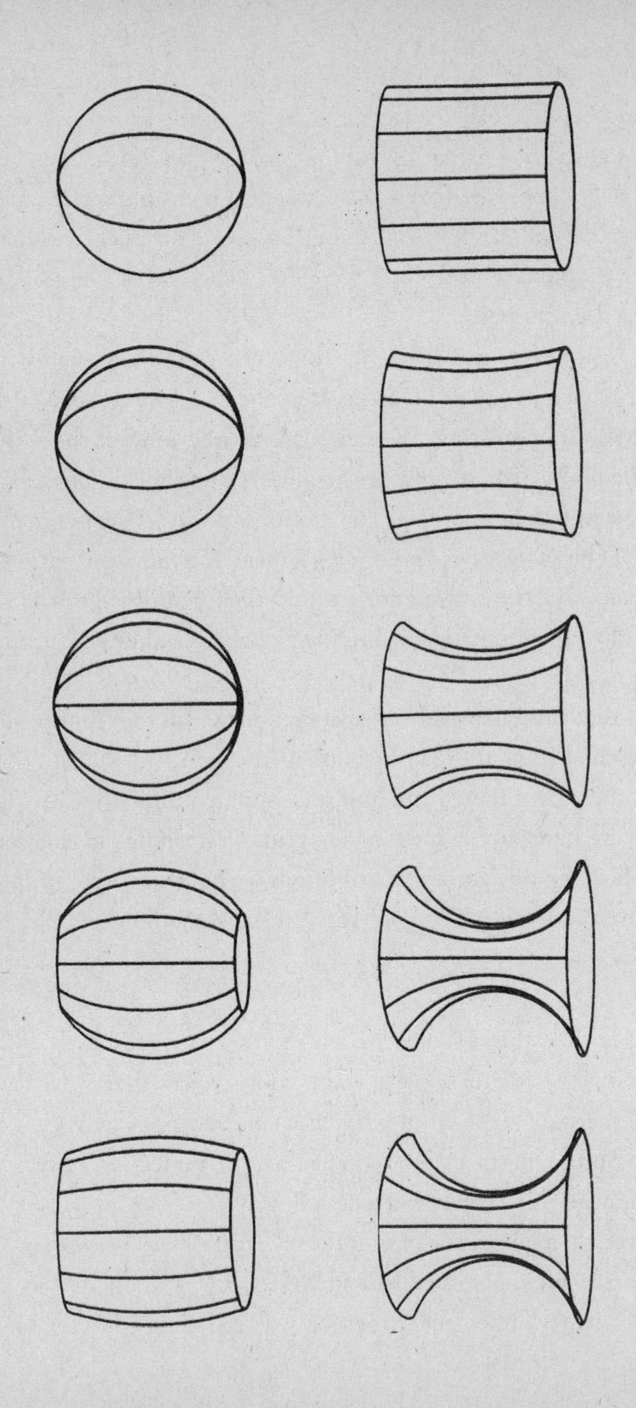

(Opposite) We can conceive a kind of space that has saddle-points in it. *Computer-graphics of the inversion of a sphere to produce a negative curvature.*

Christopher Wren after the Great Fire. The sailor trying to fix his position – longitude and latitude – off a remote shore from now on would compare his readings of the stars with those at Greenwich. The meridian of Greenwich became the fixed mark in every sailor's storm-tossed world: the meridian, and Greenwich Mean Time.

Second as an essential aid to fixing a position was the improvement of the clock. The clock became the symbol and the central problem of the age, because Newton's theories could only be put to practical use at sea if a clock could be made to keep time on a ship. The principle is simple enough. Since the sun rounds the earth in twenty-four hours, each of the 360 degrees of longitude occupies four minutes of time. A sailor who compares noon on his ship (the highest position of the sun) with noon on a clock that keeps Greenwich time therefore knows that every four minutes of difference place him one degree further away from the Greenwich meridian.

The government offered a prize of £20,000 for a time-keeper that would prove itself accurate to half a degree on a voyage of six weeks. And the London clock-makers (John Harrison, for instance) built one ingenious clock after another, designed so that their several pendulums should, between them, correct for the lurch of the ship.

These technical problems set off a burst of invention, and established the preoccupation with time that has dominated science and our daily lives ever since. A ship indeed is a kind of model of a star. How does a star ride through space, and how do we know what time it keeps? The ship is a starting point for thinking about relative time.

The clock-makers of the day were aristocrats among workmen, as the master-masons had been in the Middle Ages. It is a nice reflection that the clock as we know it, the pacemaker strapped to our pulse or the pocket dictator of modern life, had since the Middle Ages fired the skill of craftsmen too, in a leisurely way. In those days the early clock-makers wanted, not to know the time of day, but to reproduce the motions of the starry heavens.

The universe of Newton ticked on without a hitch for about two hundred years. If his ghost had come to Switzerland any time before 1900, all the clocks would have chimed hallelujah in unison. And yet, just after 1900 in Berne, not two hundred yards from the ancient clocktower, a young man came to live who was going to set them all by the ears: Albert Einstein.

Time and light first began to go awry just about this time. It was in 1881 that Albert Michelson carried out an experiment (which he repeated with Edward Morley six years later) in which he fired light in different directions, and was taken aback to find that however the apparatus moved, always he came out with the same speed of light. That was quite out of keeping with Newton's laws. And it was that small murmur at the heart of physics which first set scientists agog and questioning, about 1900.

It is not certain that the young Einstein was quite up-to-date about this. He had not been a very attentive university student. But it is certain that by the time he went to Berne he had already asked himself, years earlier as a boy in his teens, what our experience would look like seen from the point of view of light.

The answer to the question is full of paradox, and that makes it hard. And yet, as with all paradox, the hardest part is not to answer but to conceive the question. The genius of men like Newton and Einstein lies in that: they ask transparent, innocent questions which turn out to have catastrophic answers. The poet William Cowper called Newton a 'childlike sage' for that quality, and the description perfectly hits the air of surprise at the world that Einstein carried in his face. Whether he talked about riding a beam of light or falling through space, Einstein was always full of beautiful, simple illustrations of such principles, and I shall take a leaf out of his book. I go to the bottom of the clocktower, and get into the tram he used to take every day on his way to work as a clerk in the Swiss Patent Office.

The thought that Einstein had had in his teens was this: 'What would the world look like if I rode on a beam of light?' Suppose this tram were moving away from that clock on the very beam with which we see what the clock says. Then, of course, the clock would be frozen. I, the tram, this box riding on the beam of light would be fixed in time. Time would have a stop.

Let me spell that out. Suppose the clock behind me says noon when I leave. I now travel 186,000 miles away from it at the speed of light; that ought to take me one second. But the time on the clock, as I see it, still says 'noon', because it takes the beam of light from the clock exactly as long as it has taken me. So far as the clock as I see it, so far as the universe inside the tram is concerned, in keeping up with the speed of light I have cut myself off from the passage of time.

That is an extraordinary paradox. I will not go into its implications, or others that Einstein was concerned with. I will just concentrate on this point: that if I rode on a beam of light, time would suddenly come to an end for me. And that must mean that, as I approach the speed of light (which is what I am going to simulate in this tram), I am alone in my box of time and space, which is more and more departing from the norms round me.

Such paradoxes make two things clear. An obvious one: there is no universal time. But a more subtle one: that experience runs very differently for the traveller and the stay-at-home – and so for each of us on his own path. My experiences within the tram are consistent: I discover the same laws, the same relations between time, distance, speed, mass and force, that every other observer discovers. But the actual values that I get for time, distance, and so on, are not the same that the man on the pavement gets.

That is the core of the Principle of Relativity. But the obvious question is 'Well, what holds his box and mine together?' The passage of light: light is the carrier of information that binds us. And that is why the crucial experimental fact is the one that puzzled people since 1881: that when we exchange signals, then we discover that information passes between us always at the same pace. We always get the same value for the speed of light. And then naturally time and space and mass must be different for each of us, because they have to give the same laws for me here in the tram and for the man outside, consistently – yet the same value for the speed of light.

Light and the other radiations are signals that spread out from an event like ripples through the universe, and there is no way in which news of the event can move outwards faster than they do. The light or the radio wave or the X-ray is the ultimate carrier of news or messages, and forms a basic network of information which links the material universe together. Even if the message that we want to send is simply the time, we cannot get it from one place to another faster than the light or the radio wave that carries it. There is no universal time for the world, no signal from Greenwich by which we can set our watches without getting the speed of light inextricably tied up in it.

In this dichotomy, something has to give. For the path of a ray of light (like the path of a bullet) does not look the same to a casual bystander as to the man who fired it on the move. The path looks longer to the bystander; and therefore the time that the light takes on its path must seem longer to him, if he is to get the same value for its speed.

Is that real? Yes. We know enough now about cosmic and atomic processes to see that at high speeds that is true. If I were really travelling at, say, half the speed of light, then what I have been making three minutes and a little on my watch, Einstein's tram-ride, would be half a minute longer for the man on the pavement.

We will take the tram up towards the speed of light to see what the appearances look like. The relativity effect is that things change shape. (There are also changes in colour, but they are not due to relativity.) The tops of the buildings seem to bend inwards and forwards. The buildings also seem crowded together. I am travelling horizontally, so horizontal distances seem shorter; but the heights remain the same. Cars and people are distorted in the same way: thin and tall. And what is true for me looking out is true for the man outside looking in. The *Alice in Wonderland* world of relativity is symmetrical. The observer sees the tram crushed together: thin and tall.

Evidently this is an altogether different picture of the world from that which Newton had. For Newton, time and space formed an absolute framework, within which the material events of the world ran their course in imperturbable order. His is a God's eye view of the world: it looks the same to every observer, wherever he is and however he travels. By contrast, Einstein's is a man's eye view, in which what you see and what I see is relative to each of us, that is, to our place and speed. And this relativity cannot be removed. We cannot know what the world is like in itself, we can only compare what it looks like to each of us, by the practical procedure of exchanging messages. I in my tram and you in your chair can share no divine and instant view of events – we can only communicate our own views to one another. And communication is not instant; we cannot remove from it the basic time-lag of all signals, which is set by the speed of light.

The tram did not reach the speed of light. It stopped, very decently, near the Patent Office. Einstein got off, did a day's work, and often of an evening stopped at the Café Bollwerk. The work at the Patent Office was not very taxing. To tell the truth, most of the applications now look pretty idiotic: an application for an improved form of pop gun; an application for the control of alternating current, of which Einstein wrote succinctly, 'It is incorrect, inaccurate, and unclear'.

In the evenings at the Café Bollwerk he would talk a little physics with his colleagues. He would smoke cigars and drink coffee. But he was a man who thought for himself. He went to the heart of the question, which is 'How in fact do, not physicists but human beings, communicate with one another? What signals do we send from one to another? How do we reach knowledge?'

And that is the crux of all his papers, this unfolding of the heart of knowledge, almost petal by petal.

So the great paper of 1905 is not just about light or, as its title says, *The Electrodynamics of Moving Bodies.* It goes on in the same year to a postscript saying energy and mass are equivalent, $E=mc^2$. To us, it is remarkable that the first account of relativity should instantly entail a practical and devastating prediction for atomic physics. To Einstein, it is simply a part of drawing the world together; like Newton and all scientific thinkers, he was in a deep sense a unitarian. That comes from a profound insight into the processes of nature herself, but particularly into the relations between man, knowledge, nature. Physics is not events but observations. Relativity is the understanding of the world not as events but as relations.

Einstein looked back to those years with pleasure. He said to my friend Leo Szilard many years after, 'They were the happiest years of my life. Nobody expected me to lay golden eggs'. Of course, he did go on laying golden eggs: quantum effects, general relativity, field theory. With them came the confirmation of Einstein's early work, and the harvest of his predictions. In 1915 he predicted, in the General Theory of Relativity, that the gravitational field near the sun would cause a glancing ray of light to bend inwards – like a distortion of space. Two expeditions sent by the Royal Society to Brazil and the west coast of Africa tested the prediction during the eclipse on 29 May 1919. To Arthur Eddington, who was in charge of the African expedition, his first measurement of the photographs taken there always stayed in his memory as the greatest moment in his life. Fellows of the Royal Society rushed the news to one another; Eddington by telegram to the mathematician Littlewood, and Littlewood in a hasty note to Bertrand Russell,

> Dear Russell:
> Einstein's theory is completely confirmed. The predicted displacement was 1"·72 and the observed 1"·75 ± ·06.
> Yours, J.E.L.

Relativity was a fact, in the special theory and the general. $E=mc^2$ was confirmed in time, of course. Even the point about clocks running slow was singled out at last by an inexorable fate. In 1905 Einstein had written a slightly comic prescription for an ideal experiment to test it.

> If there are two synchronised clocks at A and if one of these is moved along a closed curve with constant velocity v until it returns to A, which we suppose to take t seconds, then the latter clock on arriving at A will have lost $\frac{1}{2}t\,(v/c)^2$ seconds by comparison with the clock which has remained stationary. We conclude from this that a clock fixed at the Earth's equator will run slower by a very small amount than an identical clock fixed at one of the Earth's poles.

Einstein died in 1955, fifty years after the great 1905 paper. But by then one could measure time to a thousand millionth of a second. And therefore it was possible to look at that odd proposal to 'think of two men on earth, one at the North Pole and one at the Equator. The one at the Equator is going round faster than the one at the North Pole; therefore his watch will lose'. And that is just how it turned out.

The experiment was done by a young man called H. J. Hay at Harwell. He imagined the earth squashed flat into a plate, so that the North Pole is at the centre and the equator runs round the rim. He put a radio-active clock on the rim and another at the centre of the plate and let it turn. The clocks measure time statistically by counting the number of radio-active atoms that decay. And sure enough, the clock at the rim of Hay's plate keeps time more slowly than the clock at the centre. That goes on in every spinning plate, on every turntable. At this moment, in every revolving gramophone disc, the centre is ageing faster than the rim with every turn.

Einstein was the creator of a philosophical more than a mathematical system. He had a genius for finding philosophical ideas that gave a new view of practical experience. He did not look at nature like a God but like a pathfinder, that is, a man inside the chaos of her phenomena who believed that there is a common pattern visible in them all if we look with fresh eyes. He wrote in *The World as I See It*:

> We have forgotten what features in the world of experience caused us to frame (pre-scientific) concepts, and we have great difficulty in representing the world of experience to ourselves without the spectacles of the old-established conceptual interpretation. There is the further difficulty that our language is compelled to work with words which are inseparably connected with those primitive concepts. These are the obstacles which confront us when we try to describe the essential nature of the pre-scientific concept of space.

So in a lifetime Einstein joined light to time, and time to space; energy to matter, matter to space, and space to gravitation. At the end of his life, he was still working to seek a unity between gravitation and the forces of electricity and magnetism. That is how I remember him, lecturing in the Senate House at Cambridge in an old sweater and carpet slippers with no socks, to tell us what kind of a link he was trying to find there, and what difficulties he was running his head against.

The sweater, the carpet slippers, the dislike of braces and socks, were not affectations. Einstein seemed to express, when one saw him, an article of faith from William Blake: 'Damn braces: Bless relaxes'. He was quite unconcerned about worldly success, or respectability, or conformity; most of the time he had no notion of what was expected of a man of his eminence. He hated war, and cruelty, and hypocrisy, and above all he hated dogma – except that hate is not the right word for the sense of sad revulsion that he felt; he thought hate itself a kind of dogma. He refused to become president of the state of Israel because (he explained) he had no head for human problems. It was a modest criterion, which other presidents might adopt; there would not be many survivors.

It is almost impertinent to talk of the ascent of man in the presence of two men, Newton and Einstein, who stride like gods. Of the two, Newton is the Old Testament god; it is Einstein who is the New Testament figure. He was full of humanity, pity, a sense of enormous sympathy. His vision of nature herself was that of a human being in the presence of something god-like, and that is what he always said about nature. He was fond of talking about God: 'God does not play at dice', 'God is not malicious'. Finally Niels Bohr one day said to him, 'Stop telling God what to do'. But that is not quite fair. Einstein was a man who could ask immensely simple questions. And what his life showed, and his work, is that when the answers are simple too, then you hear God thinking.

CHAPTER EIGHT

THE DRIVE FOR POWER

Revolutions are not made by fate but by men. Sometimes they are solitary men of genius. But the great revolutions in the eighteenth century were made by many lesser men banded together. What drove them was the conviction that every man is master of his own salvation.

We take it for granted now that science has a social responsibility. That idea would not have occurred to Newton or to Galileo. They thought of science as an account of the world as it is, and the only responsibility that they acknowledged was to tell the truth. The idea that science is a social enterprise is modern, and it begins at the Industrial Revolution. We are surprised that we cannot trace a social sense further back, because we nurse the illusion that the Industrial Revolution ended a golden age.

The Industrial Revolution is a long train of changes starting about 1760. It is not alone: it forms one of a triad of revolutions, of which the other two were the American Revolution that started in 1775, and the French Revolution that started in 1789. It may seem strange to put into the same packet an industrial revolution and two political revolutions. But the fact is that they were all social revolutions. The Industrial Revolution is simply the English way of making those social changes. I think of it as the English Revolution.

What makes it especially English? Obviously, it began in England. England was already the leading manufacturing nation. But the manufacture was cottage industry, and the Industrial Revolution begins in the villages. The men who make it are craftsmen: the millwright, the watchmaker, the canal builder, the blacksmith. What makes the Industrial Revolution so peculiarly English is that it is rooted in the countryside.

During the first half of the eighteenth century, in the old age of Newton and the decline of the Royal Society, England basked in a last Indian summer of village industry and the overseas trade of merchant adventurers. The summer faded. Trade grew more competitive. By the end of the century the needs of industry were harsher and more pressing. The organisation of work in the cottage was no longer productive enough. Within two generations, roughly between 1760 and 1820, the customary way of running industry changed. Before 1760, it was standard to take work to villagers in their own homes. By 1820, it was standard to bring workers into a factory and have them overseen.

We dream that the country was idyllic in the eighteenth century, a lost paradise like *The Deserted Village* that Oliver Goldsmith described in 1770.

> Sweet Auburn, loveliest village of the plain,
> Where health and plenty cheated the labouring swain.
>
> How blest is he who crowns in shades like these,
> A youth of labour with an age of ease.

That is a fable, and George Crabbe, who was a country parson and knew the villager's life at first hand, was so enraged by it that he wrote an acid, realistic poem in reply.

Yes, thus the Muses sing of happy Swains,
Because the Muses never knew their pains.

O'ercome by labour and bow'd down by time,
Feel you the barren flattery of a rhyme?

The country was a place where men worked from dawn to dark, and the labourer lived not in the sun, but in poverty and darkness. What aids there were to lighten labour were immemorial, like the mill, which was already ancient in Chaucer's time. The Industrial Revolution began with such machines; the millwrights were the engineers of the coming age. James Brindley of Staffordshire started his self-made career in 1733 by working at mill wheels, at the age of seventeen, having been born poor in a village.

Brindley's improvements were practical: to sharpen and step up the performance of the water wheel as a machine. It was the first multi-purpose machine for the new industries. Brindley worked, for example, to improve the grinding of flints, which were used in the rising pottery industry.

Yet there was a bigger movement in the air by 1750. Water had become the engineers' element, and men like Brindley were possessed by it. Water was gushing and fanning out all over the countryside. It was not simply a source of power, it was a new wave of movement. James Brindley was a pioneer in the art of building canals or, as it was then called, 'navigation'. (It was because Brindley could not spell the word 'navigator' that workmen who dig trenches or canals are still called 'navvies'.)

Brindley had begun on his own account, out of interest, to survey the waterways that he travelled as he went about his engineering projects for mills and mines. The Duke of Bridgewater then got him to build a canal to carry coal from the Duke's pits at Worsley to the rising town of Manchester. It was a prodigious design, as a letter to the Manchester Mercury recorded in 1763.

> I have lately been viewing the artificial wonders of London and natural wonders of the Peak, but none of them gave me so much pleasure as the Duke of Bridgewater's navigation in this country. His projector, the ingenious Mr Brindley, has indeed made such improvements in this way as are truly astonishing. At Barton Bridge, he has erected a navigable canal in the air; for it is as high as the tree-tops. Whilst I was surveying it with a mixture of wonder and delight, four barges passed me in the space of about three minutes, two of them being chained together, and dragged by two horses, who went on the terrace of the canal, whereon I durst hardly venture ... to walk, as I almost trembled to behold the large River Irwell underneath me. Where Cornebrooke comes athwart the Duke's navigation ... about a mile from Manchester, the Duke's agents have made a wharf and are selling coals at three pence halfpenny per basket ... Next summer they intend to land them in (Manchester).

Brindley went on to connect Manchester with Liverpool in an even bolder manner, and in all laid out almost four hundred miles of canals in a network all over England.

Two things are outstanding in the creation of the English system of canals, and they characterise all the Industrial Revolution. One is that the men who made the revolution were practical men. Like Brindley, they often had little education, and in fact school education as it then was could only dull an inventive mind. The grammar schools legally could only teach the classical subjects for which they had been founded. The universities also (there were only two, at Oxford and Cambridge) took little interest in modern or scientific studies; and they were closed to those who did not conform to the Church of England.

The other outstanding feature is that the new inventions were for everyday use. The canals were arteries of communication: they were not made to carry pleasure boats, but barges. And the barges were not made to carry luxuries, but pots and pans and bales of cloth, boxes of ribbon, and all the common things that people buy by the pennyworth. These things had been manufactured in villages which were growing into towns now, away from London; it was a country-wide trade.

Technology in England was for use, up and down the country, far from the capital. And that is exactly what technology was *not* in the dark confines of the courts of Europe. For example, the French and the Swiss were quite as clever as the English (and much more ingenious) in making scientific playthings. But they lavished that clockwork brilliance on making toys for rich or royal patrons. The automata on which they spent years are to this day the most exquisite in the flow of movement that have ever been made. The French were the inventors of automation: that is, of the idea of making each step in a sequence of movements control the next. Even the modern control of machines by punched cards had already been devised by Joseph Marie Jacquard about 1800, for the silk-weaving looms of Lyons, and languished in such luxury employment.

Fine skill of this kind could advance a man in France before the revolution. A watchmaker, Pierre Caron, who invented a new watch escapement and pleased Queen Marie Antoinette, prospered at court and became Count Beaumarchais. He had musical and literary talent, too, and he later wrote a play on which Mozart based his opera *The Marriage of Figaro*. Although a comedy seems an unlikely source book of social history, the intrigues in and about the play reveal how talent fared at the courts of Europe.

At first sight *The Marriage of Figaro* looks like a French puppet play, humming with secret machinations. But the fact is that it is an early storm signal of the revolution. Beaumarchais had a fine political nose for what was cooking, and supped with a long spoon. He was employed by the royal ministers in several double-edged deals, and on their behalf in fact was involved in a secret arms deal with the American revolutionaries to help them fight the English. The King might believe that he was playing at Machiavelli, and that he could keep such contrivances of policy for export only. But Beaumarchais was more sensitive and more astute, and could smell the revolution coming home. And the message he put into the character of Figaro, the servant, is revolutionary.

> Bravo, Signor Padrone –
> Now I'm beginning to understand all this mystery, and to appreciate your most generous intentions. The King appoints you Ambassador in London, I go as courier and my Susanna as confidential attachée. No, I'm hanged if she does – Figaro knows better.

Mozart's famous aria, 'Count, little Count, you may go dancing, but I'll play the tune' (*Se vuol ballare, Signor Contino...*) is a challenge. In Beaumarchais's words it runs:

> No, my lord Count, you shan't have her, you shan't. Because you are a great lord, you think you're a great genius. Nobility, wealth, honours, emoluments! They all make a man so proud! What have you done to earn so many advantages? You took the trouble to be born, nothing more. Apart from that, you're rather a common type.
>
> A public debate started on the nature of wealth, and since one needn't own something in order to argue about it, being

> in fact penniless, I wrote on the value of money and interest. Immediately, I found myself looking at... the drawbridge of a prison ... Printed nonsense is dangerous only in countries where its free circulation is hampered; without the right to criticise, praise and approval are worthless.

That was what was going on under the courtly pattern of French society, as formal as the garden of the Château at Villandry.

It seems inconceivable now that the garden scene in *The Marriage of Figaro*, the aria in which Figaro dubs his master 'Signor Contino', little Count, should in their time have been thought revolutionary. But consider when they were written. Beaumarchais finished the play of *The Marriage of Figaro* about 1780. It took him four years of struggle against a host of censors, above all Louis XVI himself, to get a performance. When it was performed, it was a scandal over Europe. Mozart was able to show it in Vienna by turning it into an opera. Mozart was thirty then; that was in 1786. And three years later, in 1789 – the French Revolution.

Was Louis XVI toppled from his throne and beheaded because of *The Marriage of Figaro*? Of course not. Satire is not a social dynamite. But it is a social indicator: it shows that new men are knocking at the door. What made Napoleon call the last act of the play 'the revolution in action'? It was Beaumarchais himself, in the person of Figaro, pointing to the Count and saying, 'Because you are a great nobleman, you think you are a great genius. You have taken trouble with nothing, except to be born'.

Beaumarchais represented a different aristocracy, of working talent: the watchmakers in his age, the masons in the past, the printers. What excited Mozart about the play? The revolutionary ardour, which to him was represented by the movement of Freemasons to which he belonged, and which he glorified in *The Magic Flute*. (Freemasonry was then a rising and secret society whose undertone was anti-establishment and anti-clerical, and because Mozart was known to be a member it was difficult to get a priest to come to his deathbed in 1791.) Or think of the greatest Freemason of them all in that age, the printer Benjamin Franklin. He was American emissary in France at the Court of Louis XVI in 1784 when *The Marriage of Figaro* was first performed. And he more than anyone else represents those forward looking, forceful, confident, thrusting, marching men who made the new age.

For one thing, Benjamin Franklin had such marvellous luck. When he went to present his credentials to the French Court in 1778, it turned out at the last moment that the wig and formal clothes were too small for him. So he boldly went in his own hair, and was instantly hailed as the child of nature from the backwoods.

All his actions have the stamp of a man who knows his mind, and knows the words to speak it. He published an annual, *Poor Richard's Almanack*, which is full of the raw material for future proverbs: 'Hunger never saw bad bread.' 'If you want to know the value of money, try to borrow some.' Franklin wrote of it:

> In 1732 I first published my Almanac ... It was continued by me about 25 years ... I endeavoured to make it both entertaining and useful, and it accordingly came to be in such demand that I reaped considerable profit from it; vending annually near ten thousand ... scarce any neighbourhood in the province being without it. I considered it as a proper vehicle for conveying instruction among the common people, who bought scarcely any other books.

To those who doubted the use of new inventions (the occasion was the first hydrogen balloon ascent in Paris in 1783) Franklin replied, 'What

(Right) A lightning conductor dating from Franklin's day.

is the use of a new-born baby?' His character is condensed in the answer, optimistic, down to earth, pithy, and memorable enough to be used again by Michael Faraday, a greater scientist, in the next century. Franklin was alive to how things were said. He made the first pair of bifocal spectacles for himself by sawing his lenses in half, because he could not follow French at Court unless he could watch the speaker's expression.

Men like Franklin had a passion for rational knowledge. Looking at the mountain of neat achievements scattered through his life, the pamphlets, the cartoons, the printer's stamps, we are struck by the spread and richness of his inventive mind. The scientific entertainment of the day was electricity. Franklin loved fun (he was a rather improper man), yet he took electricity seriously; he recognised it as a force in nature. He proposed that lightning is electric, and in 1752 he proved it – how would a man like Franklin prove it? – by hanging a key from a kite in a thunderstorm. Being Franklin, his luck held; the experiment did not kill him, only those who copied it. Of course, he turned his experiment into a practical invention, the lightning conductor; and made it illuminate the theory of electricity too by arguing that all electricity is of one kind and not, as was then thought, two different fluids.

There is a footnote to the invention of the lightning conductor to remind us again that social history hides in unexpected places. Franklin reasoned, rightly, that the lightning conductor would work best with a sharp end. This was disputed by some scientists, who argued for a rounded end, and the Royal Society in England had to arbitrate. However, the argument was settled at a more primitive and elevated level: King George III, in a rage against the American revolution, fitted rounded ends to the lightning conductors on royal buildings. Political interference with science is usually tragic; it is happy to have a comic instance that rivals the war in *Gulliver's Travels* between 'the two great Empires of *Lilliput* and *Blefuscu*' that opened their breakfast egg at the sharp or the rounded end.

Franklin and his friends lived science; it was constantly in their thoughts and just as constantly in their hands. The understanding of nature to them was an intensely practical pleasure. These were men in society: Franklin was a political man, whether he printed paper money or his endless racy pamphlets. And his politics were as downright as his experiments. He changed the florid opening of the Declaration of Independence to read with simple confidence, 'We hold these truths to be *self-evident*, that all men are created equal'. When war between England and the American revolutionaries broke out, he wrote openly to an English politician who had been his friend, in words charged with fire:

> You have begun to burn our towns. Look upon your hands! They are stained with the blood of your relations.

Ironmasters like John Wilkinson minted their own wage tokens, with their own unroyal faces on them.
A Wilkinson token, 1788.

The red glow has become the picture of the new age in England – in the sermons of John Wesley, and in the furnace sky of the Industrial Revolution, such as the fiery landscape of Abbeydale in Yorkshire, an early centre for new processes in making iron and steel. The masters of industry were the ironmasters: powerful, more than life-size, demonic figures, whom governments suspected, rightly, of really believing that all men are created equal. The working men in the north and the west were no longer farm labourers, they were now an industrial community. They had to be paid in coin, not in kind. Governments in London were remote from all this. They refused to mint enough small change, so ironmasters like John Wilkinson minted their own wage tokens, with their own unroyal faces on them. Alarm in London: was this a Republican plot? No, it was not a plot. But it is true that radical inventions came out of radical brains. The first model of an iron bridge to be exhibited in London was proposed by Tom Paine, a firebrand in America and in England, protagonist of *The Rights of Man*.

Meanwhile, cast iron was already being used in revolutionary ways by the ironmasters like John Wilkinson. He built the first iron boat in 1787, and boasted that it would carry his coffin when he died. And he was buried in an iron coffin in 1808. Of course, the boat sailed under an iron bridge; Wilkinson had helped to build that in 1779 at a nearby Shropshire town that is still called Ironbridge.

Did the architecture of iron really rival the architecture of the cathedrals? It did. This was a heroic age. Thomas Telford felt that, spanning the landscape with iron. He was born a poor shepherd, then worked as a journeyman mason, and on his own initiative became an engineer of roads and canals, and a friend of poets. His great aqueduct that carries the Llangollen canal over the river Dee shows him to have been a master of cast iron on the grand scale. The monuments of the Industrial Revolution have a Roman grandeur, the grandeur of Republican men.

The men who made the Industrial Revolution are usually pictured as hardfaced businessmen with no other motive than self-interest. That is certainly wrong. For one thing, many of them were inventors who had come into business that way. And for another, a majority of them were not members of the Church of England but belonged to a puritan tradition in the Unitarian and similar movements. John Wilkinson was much under the influence of his brother-in-law Joseph Priestley, later famous as a chemist, but who was a Unitarian minister and was probably the pioneer of the principle, 'the greatest happiness of the greatest number'.

Joseph Priestley, in turn, was scientific adviser to Josiah Wedgwood. Now Wedgwood we usually think of as a man who made marvellous tableware for aristocracy and royalty: and so he did, on rare occasions, when he got the commission. For example, in 1774 he made a service of nearly a thousand highly decorated pieces for Catherine the Great of Russia, which cost over £2000

– a great deal of money in the coin of that day. But the base of that tableware was his own pottery, creamware; and in fact all the thousand pieces, undecorated, cost less than £50, yet looked and handled like Catherine the Great's in every way except for the hand-painted idylls. The creamware which made Wedgwood famous and prosperous was not porcelain, but a white earthenware pottery for common use. That is what the man in the street could buy, at about a shilling a piece. And in time that is what transformed the kitchens of the working class in the Industrial Revolution.

Wedgwood was an extraordinary man: inventive, of course, in his own trade, and also in the scientific techniques that might make his trade more exact. He invented a way of measuring the high temperatures in the kiln by means of a sort of sliding scale of expansion in which a clay test-piece moved. Measuring high temperatures is an ancient and difficult problem in the manufacture of ceramics and metals, and it is fitting (as things went then) that Wedgwood was elected to the Royal Society.

Josiah Wedgwood was no exception; there were dozens of men like him. Indeed, he belonged to a group of about a dozen men, the Lunar Society of Birmingham (Birmingham was then still a scattered group of industrial villages), who gave themselves the name because they met near the full moon. This was so that people like Wedgwood, who came from a distance to Birmingham, should be able to travel safely over wretched roads that were dangerous on dark nights.

But Wedgwood was not the most important industrialist there: that was Matthew Boulton, who brought James Watt to Birmingham because there they could build the steam engine. Boulton was fond of talking about measurement; he said that nature had destined him to be an engineer by having him born in the year 1728, because that is the number of cubic inches in a cubic foot. Medicine was important in that group also, for there were new and important advances being made. Dr William Withering discovered the use of digitalis in Birmingham. One of the doctors who has remained famous, who belonged to the Lunar Society, was Erasmus Darwin, the grandfather of Charles Darwin. The other grandfather? Josiah Wedgwood.

Societies like the Lunar Society represent the sense of the makers of the Industrial Revolution (that very English sense) that they had a social responsibility. I call it an English sense, though in fact that is not quite fair; the Lunar Society was much influenced by Benjamin Franklin and by other Americans associated with it. What ran through it was a simple faith: the good life is *more* than material decency, but the good life must be *based* on material decency.

It took a hundred years before the ideals of the Lunar Society became reality in Victorian England. When it did come, the reality seemed commonplace, even comic, like a Victorian picture postcard. It is comic to think that cotton underwear and soap could work a transformation in the lives of the poor. Yet these simple things – coal in an iron range, glass in the windows, a choice of food – were a wonderful rise in the standard of life and health. By our standards, the industrial towns were slums, but to the people who had come from a cottage, a house in a terrace was a liberation from hunger, from dirt, and from disease; it offered a new wealth of choice. The bedroom with the text on the wall seems funny and pathetic to us, but for the working class wife it was the first experience of private decency. Probably the iron bedstead saved more women from childbed fever than the doctor's black bag, which was itself a medical innovation.

These benefits came from mass production in factories. And the factory system was ghastly; the schoolbooks are right about that. But it was ghastly in the old traditional way. Mines and workshops had been dank, crowded and tyrannical long before the Industrial Revolution.

The factories simply carried on as village industry had always done, with a heartless contempt for those who worked in them.

Pollution from the factories was not new either. Again, it was the tradition of the mine and the workshop, which had always fouled their environment. We think of pollution as a modern blight, but it is not. It is another expression of the squalid indifference to health and decency that in past centuries had made the Plague a yearly visitation.

The new evil that made the factory ghastly was different: it was the domination of men by the pace of the machines. The workers for the first time were driven by an inhuman clockwork: the power first of water and then of steam. It seems insane to us (it was insane) that manufacturers should be intoxicated by the gush of power that spurted from the factory boiler without a stop. A new ethic was preached in which the cardinal sin was not cruelty or vice, but idleness. Even the Sunday schools warned children that

> *Satan* finds some Mischief still
> For idle Hands to do.

The change in the scale of time in the factories was ghastly and destructive. But the change in the scale of power opened the future. Matthew Boulton of the Lunar Society, for example, built a factory which was a showplace, because Boulton's kind of metalwork depended on the skill of craftsmen. Here James Watt came to build the sun-god of all power, the steam engine, because only here was he able to find the standards of accuracy needed to make the engine steam-tight.

In 1776 Matthew Boulton was very excited about his new partnership with James Watt to build the steam engine. When James Boswell, the biographer, came to see Boulton that year, he said to him grandly, 'I sell here, sir, what all the world desires to have – power'. It is a lovely phrase. But it is also true.

Power is a new preoccupation, in a sense a new idea, in science. The Industrial Revolution, the English revolution, turned out to be the great discoverer of power. Sources of energy were sought in nature: wind, sun, water, steam, coal. And a question suddenly became concrete: Why are they all one? What relation exists between them? That had never been asked before. Until then science had been entirely concerned with exploring nature as she is. But now the modern conception of transforming nature in order to obtain power from her, and of changing one form of power into another, had come up to the leading edge of science. In particular, it grew clear that heat is a form of energy, and is converted into other forms at a fixed rate of exchange. In 1824 Sadi Carnot, a French engineer, looking at steam engines, wrote a treatise on what he called 'la puissance motrice du feu', in which he founded, in essence, the science of thermodynamics – the dynamics of heat. Energy had become a central concept in science; and the main concern in science now was the unity of nature, of which energy is the core.

And it was a main concern not only in science. You see it equally in the arts, and the surprise is there. While this is going on, what is going on in literature? The uprush of romantic poetry round about the year 1800. How could the romantic poets be interested in industry? Very simply: the new concept of nature as the carrier of energy took them by storm. They loved the word 'storm' as a synonym for energy, in phrases like *Sturm und Drang*, 'storm and thrust'. The climax of Samuel Taylor Coleridge's *Rime of the Ancient Mariner* is introduced by a storm that breaks the deadly calm and releases life again.

> The upper air burst into life!
> And a hundred fire-flags sheen,
> To and fro they were hurried about!

And to and fro, and in and out,
The wan stars danced between.

The loud wind never reached the ship,
Yet now the ship moved on!
Beneath the lightning and the Moon
The dead men gave a groan.

A young German philosopher, Friedrich von Schelling, just at this time in 1799, started a new form of philosophy which has remained powerful in Germany, *Naturphilosophie* – philosophy of nature. From him Coleridge brought it to England. The Lake Poets had it from Coleridge, and the Wedgwoods, who were friends of Coleridge's and indeed supported him with an annuity. Poets and painters were suddenly captured by the idea that nature is the fountain of power, whose different forms are all expressions of the same central force, namely energy.

And not only nature. Romantic poetry says in the plainest way that man himself is the carrier of a divine, at least a natural, energy. The Industrial Revolution created freedom (in practice) for men who wanted to fulfil what they had in them – a concept inconceivable a hundred years earlier. But hand in hand, romantic thought inspired those men to make of their freedom a new sense of personality in nature. It was said best of all by the greatest of the romantic poets, William Blake, very simply: 'Energy is Eternal Delight'.

The key word is 'delight', the key concept is 'liberation' – a sense of fun as a human right. Naturally, the marching men of the age expressed the impulse in invention. So they produced a bottomless horn of plenty of eccentric ideas to delight the Saturday evenings of the working family. (To this day, most of the applications that lumber the patent offices are

(*Opposite*) So they produced a bottomless horn of plenty of eccentric ideas to delight the Saturday evenings of the working family.
A patent elevator-platform.

slightly mad, like the inventors themselves.) We could build an avenue from here to the moon lined with these lunacies, and it would be just about as pointless and yet as high-spirited as getting to the moon. Consider, for example, the idea of the zoetrope, a circular machine for animating a Victorian comic strip by flashing the pictures past the eye one after another. It is quite as exciting as an evening at the cinema, and comes to the point rather quicker. Or the automatic orchestra, which has the advantage of a very small repertoire. All of it is packed with homespun vigour which has not heard of good taste, and is absolutely self-made. Every pointless invention for the household, like the mechanical vegetable chopper, is matched by another superb one, like the telephone. And finally, at the end of the avenue of pleasure, we should certainly put the machine that is the essence of machineness: it does nothing at all!

The men who made the wild inventions and the grand ones came from the same mould. Think of the invention that rounded out the Industrial Revolution as the canals had begun it: the railways. They were made possible by Richard Trevithick, who was a Cornish blacksmith and a wrestler and a strong man. He turned the steam engine into a mobile power pack by changing Watt's beam engine into a high-pressure engine. It was a life-giving act, which opened a blood-stream of communication for the world, and made England the heart of it.

We are still in the middle of the Industrial Revolution; we had better be, for we have many things to put right in it. But it has made our world richer, smaller, and for the first time ours. And I mean that literally: our world, everybody's world.

From its earliest beginnings, when it was still dependent on water power, the Industrial Revolution was terribly cruel to those whose lives and livelihood it overturned. Revolutions are – it is their nature, because by definition revolutions move too fast for those whom they strike. Yet it became in time a social revolution and established that social equality, the equality of rights, above all intellectual equality, on which we depend. Where would a man like me be, where would you be, if we had been born before 1800? We still live in the middle of the Industrial Revolution and find it hard to see its implications, but the future will say of it that in the ascent of man it is a step, a stride, as powerful as the Renaissance. The Renaissance established the dignity of man. The Industrial Revolution established the unity of nature.

That was done by scientists and romantic poets who saw that the wind and the sea and the stream and the steam and the coal are all created by the heat of the sun, and that heat itself is a form of energy. A good many men thought of that, but it was established above all by one man, James Prescott Joule of Manchester. He was born in 1818, and from the age of twenty spent his life in the delicate detail of experiments to determine the mechanical equivalent of heat – that is, to establish the exact rate of exchange at which mechanical energy is turned into heat. And since that sounds a very solemn and boring undertaking, I must tell a funny story about him.

In the summer of 1847, the young William Thomson (later to be the great Lord Kelvin, the panjandrum of British science) was walking – where does a British gentleman walk in the Alps? – from Chamonix to Mont Blanc. And there he met – whom does a British gentleman meet in the Alps? – a British eccentric: James Joule, carrying an enormous thermometer and accompanied at a little distance by his wife in a carriage. All his life, Joule had wanted to demonstrate that water, when it falls through 778 feet, rises one degree Fahrenheit in temperature. Now on his honeymoon he could decently visit Chamonix (rather as

American couples go to Niagara Falls) and let nature run the experiment for him. The waterfall here is ideal. It is not all of 778 feet, but he would get about half a degree Fahrenheit. As a footnote, I should say that he did not – of course – actually succeed; alas, the waterfall is too broken by spray for the experiment to work.

The story of the British gentlemen at their scientific eccentricities is not irrelevant. It was such men who made nature romantic; the Romantic Movement in poetry came step by step with them. We see it in poets like Goethe (who was also a scientist) and in musicians like Beethoven. We see it first of all in Wordsworth: the sight of nature as a new quickening of the spirit because the unity in it was immediate to the heart and mind. Wordsworth had come through the Alps in 1790 when he had been drawn to the Continent by the French Revolution. And in 1798 he said, in *Tintern Abbey*, what could not be said better.

> For nature then ...
> To me was all in all – I cannot paint
> What then I was. The sounding cataract
> Haunted me like a passion.

'Nature then to me was all in all.' Joule never said it as well as that. But he did say, 'The grand agents of nature are indestructible', and he meant the same things.

CHAPTER NINE

THE LADDER OF CREATION

The theory of evolution by natural selection was put forward in the 1850s independently by two men. One was Charles Darwin; the other was Alfred Russel Wallace. Both men had some scientific background, of course, but at heart both men were naturalists. Darwin had been a medical student at Edinburgh University for two years, before his father who was a wealthy doctor proposed that he might become a clergyman and sent him to Cambridge. Wallace, whose parents were poor and who had left school at fourteen, had followed courses at Working Men's Institutes in London and Leicester as a surveyor's apprentice and pupil teacher.

The fact is that there are two traditions of explanation that march side by side in the ascent of man. One is the analysis of the physical structure of the world. The other is the study of the processes of life: their delicacy, their diversity, the wavering cycles from life to death in the individual and in the species. And these traditions do not come together until the theory of evolution; because until then there is a paradox which cannot be resolved, which cannot be begun, about life.

The paradox of the life sciences, which makes them different in kind from physical science, is in the detail of nature everywhere. We see it about us in the birds, the trees, the grass, the snails, in every

The theory of evolution was conceived twice by two men living at the same time in the same culture.
Charles Darwin.

living thing. It is this. The manifestations of life, its expressions, its forms, are so diverse that they must contain a large element of the accidental. And yet the nature of life is so uniform that it must be constrained by many necessities.

So it is not surprising that biology as we understand it begins with naturalists in the eighteenth and nineteenth centuries: observers of the countryside, bird-watchers, clergymen, doctors, gentlemen of leisure in country houses. I am tempted to call them, simply, 'gentlemen in Victorian England', because it cannot be an accident that the theory of evolution is conceived twice by two men living at the same time in the same culture – the culture of Queen Victoria in England.

Charles Darwin was in his early twenties when the Admiralty was about to send out a survey ship called the *Beagle* to map the coast of South America, and he was offered the unpaid post of naturalist. He owed the invitation to the professor of botany who had befriended him at Cambridge, though Darwin had not been excited by botany there but by collecting beetles.

> I will give a proof of my zeal: one day, on tearing off some old bark, I saw two rare beetles, and seized one in each hand; then I saw a third and new kind, which I could not bear to lose, so that I popped the one which I held in my right hand into my mouth.

Darwin's father opposed his going, and the captain of the *Beagle* did not like the shape of his nose, but Darwin's Wedgwood uncle spoke up for him and he went. The *Beagle* set sail on 27 December 1831.

The five years that he spent on the ship transformed Darwin. He had been a sympathetic, subtle observer of birds, flowers, life in his own countryside; now South America exploded all that for him into a passion. He came home convinced that species are taken in different directions when they are isolated from one another; species are not immutable. But when he came back he could not think of any mechanism that drove them apart. That was in 1836.

When Darwin did hit on an explanation for the evolution of species two years later, he was most reluctant to publish it. He might have put it off all his life if a very different kind of man had not also followed almost exactly the same steps of experience and thought that moved Darwin, and arrived at the same theory. He is the forgotten and yet the vital character, a sort of man from Porlock in reverse, in the theory of evolution by natural selection.

His name was Alfred Russel Wallace, a giant of a man with a Dickensian family history as comic as Darwin's was stuffy. At that time, in 1836, Wallace was a boy in his teens; he was born in 1823, and that makes him fourteen years younger than Darwin. Wallace's life was not easy even then.

> Had my father been a moderately rich man ... my whole life would have been differently shaped, and though I should, no doubt, have given some attention to science, it seems very unlikely that I should have ever undertaken ... a journey to the almost unknown forests of the Amazon in order to observe nature and make a living by collecting.

So Wallace wrote about his early life, when he had had to find a way to earn his living in the English provinces. He took up the profession of land-surveying, which did not require a university education, and which his older brother could teach him. His brother died in 1846 from a chill he caught travelling home in an open third-class carriage from a meeting of a Royal Commission committee on rival railway firms.

Evidently it was an open-air life, and Wallace became interested in plants and insects. When he was working at Leicester, he met a man with the same interests who was rather better educated. His new friend astonished Wallace by telling him that he had collected several hundred different species of beetles in the neighbourhood of Leicester, and that there were more to be discovered.

> If I had been asked before how many different kinds of beetles were to be found in any small district near a town, I should probably have guessed fifty ... I now learnt ... that there were probably a thousand different kinds within ten miles.

It was a revelation to Wallace, and it shaped his life and his friend's. The friend was Henry Bates, who later did famous work on mimicry among insects.

Meanwhile the young man had to make a living. Fortunately, it was a good time for a land-surveyor, because the railway adventurers of the 1840s needed him. Wallace was employed to survey a possible route for a line in the Neath Valley in South Wales. He was a conscientious technician, as his brother had been and as Victorians were. But he suspected rightly that he was a pawn in a power game. Most of the surveys were only meant to establish a claim against some other railway robber baron. Wallace calculated that only a tenth of the lines surveyed that year were ever built.

The Welsh countryside was a delight to the Sunday naturalist, as happy in his science as a Sunday painter is in his art. Now Wallace observed and collected for himself, with a growing excitement in the variety of nature that affectionately remained in his memory all his life.

> Even when we were busy I had Sundays perfectly free, and used then to take long walks over the mountains with my collecting

> box, which I brought home full of treasures ... At such times I experienced the joy which every discovery of a new form of life gives to the lover of nature, almost equal to those raptures which I afterwards felt at every capture of new butterflies on the Amazon.

Wallace found a cave on one of his weekends where the river ran underground, and decided then and there to camp overnight. It was as if unconsciously he was already preparing himself for life in the wild.

> We wanted for once to try sleeping out-of-doors, with no shelter or bed but what nature provided ... I think we had determined purposely to make no preparation, but to camp out just as if we had come accidentally to the place in an unknown country, and had been compelled to sleep there.

In fact he hardly slept at all.

When he was twenty-five, Wallace decided to become a full-time naturalist. It was an odd Victorian profession. It meant that he would have to keep himself by collecting specimens in foreign parts to sell to museums and collectors in England. And Bates would come with him. So the two of them set off in 1848 with £100 between them. They sailed to South America, and then a thousand miles up the Amazon to the city of Manaus, where the Amazon is joined by the Rio Negro.

Wallace had hardly been further than Wales, but he was not overawed by the exotic. From the moment of arrival, his comments were firm and self-assured. For example, on the subject of vultures, he records his thoughts in his *Narrative of Travels on the Amazon and Rio Negro* five years later.

> The common black vultures were abundant, but were rather put to it for food, being obliged to eat palm-fruits in the forest when they could find nothing else.
>
> I am convinced, from repeated observations, that the vultures depend entirely on sight, and not at all on smell, in seeking out their food.

The friends separated at Manaus, and Wallace set off up the Rio Negro. He was looking for places that had not been much explored by earlier naturalists; for if he was going to make a living by collecting, he needed to find specimens of unknown or at least of rare species. The river was swollen with rain, so that Wallace and his Indians were able to take their canoe right into the forest. The trees hung low over the water. Wallace for once was awed by the gloom, but he was also elated by the variety in the forest, and he speculated how it might look from the air.

> What we may fairly allow of tropical vegetation is, that there is a much greater number of species, and a greater variety of forms, than in the temperate zones.
>
> Perhaps no country in the world contains such an amount of vegetable matter on its surface as the valley of the Amazon. Its entire extent, with the exception of some very small portions, is covered with one dense and lofty primeval forest, the most extensive and unbroken which exists upon the earth.
>
> The whole glory of these forests could only be seen by sailing gently in a balloon over the undulating flowery surface above: such a treat is perhaps reserved for the traveller of a future age.

He was excited and frightened when for the first time he went into a native Indian village; but it is characteristic of Wallace that his lasting feeling was pleasure.

> The ... most unexpected sensation of surprise and delight was my first meeting and living with a man in a state of nature – with absolute uncontaminated savages! ... They were all going about their own work or pleasure which had nothing to do with white men or their ways; they walked with the free step of the independent forest-dweller, and ... paid no attention whatever to us, mere strangers of an alien race.
>
> In every detail they were original and self-sustaining, as are the wild animals of the forests, absolutely independent of civilisation, and who could and did live their lives in their own way, as they had done for countless generations before America was discovered.

It turned out that the Indians were not fierce but helpful. Wallace drew them into the business of collecting specimens.

> During the time I remained here (forty days), I procured at least forty species of butterflies quite new to me, besides a considerable collection of other orders.
>
> One day I had brought me a curious little alligator of a rare species, with numerous ridges and conical tubercles, *Caiman gibbus*, which I skinned and stuffed, much to the amusement of the Indians, half a dozen of whom gazed intently at the operation.

Sooner or later, amid the pleasures and the labours of the forest, the burning question began to flicker in Wallace's acute mind. How had all this variety come about, so alike in design and yet so changeable in detail? Like Darwin, Wallace was struck by the differences between neighbouring species, and like Darwin he began to wonder how they had come to develop so differently.

> There is no part of natural history more interesting or instructive than the study of the geographical distribution of animals.
>
> Places not more than fifty or a hundred miles apart often have species of insects and birds at the one, which are not found at the other. There must be some boundary which determines the range of each species; some external peculiarity to mark the line which each one does not pass.

He was always attracted by problems in geography. Later, when he worked in the Malay archipelago, he showed that the animals on the western islands resemble species from Asia, and on the eastern islands from Australia: the line that divides them is still called the Wallace line.

Wallace was as acute an observer of men as of nature, and with the same interest in the origin of differences. In an age in which Victorians called the people of the Amazon 'savages', he had a rare sympathy with their culture. He understood what language, what invention, what custom meant to them. He was perhaps the first person to seize the fact that the cultural distance between their civilisation and ours is much shorter than we think. After he conceived the principle of natural selection, that seemed not only true but biologically obvious.

> Natural selection could only have endowed savage man with a brain a few degrees superior to that of an ape, whereas he actually possesses one very little inferior to that of a philosopher. With our advent there had come into existence a being in whom that subtle force we term 'mind' became of far more importance than mere bodily structure.

He was steadfast in his regard for the Indians, and he wrote an idyllic account of their life when he stayed in the village of Javíta in 1851. At this point, Wallace's journal breaks into poetry – well, into verse.

There is an Indian village; all around,
The dark, eternal, boundless forest spreads
Its varied foliage.

Here I dwelt awhile, the one white man
Among perhaps two hundred living souls.

Each day some labour calls them. Now they go
To fell the forest's pride, or in canoe
With hook, and spear, and arrow, to catch fish;

A palm-tree's spreading leaves supply a thatch
Impervious to the winter's storms and rain.

The women dig the mandiocca root,
And with much labour make of it their bread.

And all each morn and eve wash in the stream,
And sport like mermaids in the sparkling wave.

The children of small growth are naked, and
The boys and men wear but a narrow cloth.
How I delight to see those naked boys!
Their well-form'd limbs, their bright, smooth,
red-brown skin,
And every motion full of grace and health;
And as they run, and race, and shout, and leap,
Or swim and dive beneath the rapid stream,

I pity English boys; their active limbs
Cramp'd and confined in tightly-fitting clothes;

> But how much more I pity English maids,
> Their waist, and chest, and bosom all confined
> By that vile torturing instrument called stays!
>
> I'd be an Indian here, and live content
> To fish, and hunt, and paddle my canoe,
> And see my children grow, like young wild fawns,
> In health of body and in peace of mind,
> Rich without wealth, and happy without gold!

The sympathy is different from the feelings that South American Indians aroused in Charles Darwin. When Darwin met the natives of Tierra del Fuego he was horrified: that is clear from his own words and from the drawings in his book on *The Voyage of the Beagle*. No doubt the ferocious climate had an influence on the customs of the Fuegians. But nineteenth-century photographs show that they did not look as beastly as they seemed to Darwin. On his voyage home, Darwin had published a pamphlet with the captain of the *Beagle* at Cape Town to recommend the work that missionaries were doing to change the life of savages.

Wallace spent four years in the Amazon basin; then he packed his collections and started home.

> The fever and ague now attacked me again, and I passed several days very uncomfortably. We had almost constant rains; and to attend to my numerous birds and animals was a great annoyance, owing to the crowded state of the canoe, and the impossibility of properly cleaning them during the rain. Some died almost every day, and I often wished I had nothing whatever to do with them, though, having once taken them in hand, I determined to persevere.

Out of a hundred live animals which I had purchased or had had given to me, there now only remained thirty-four.

The voyage home went badly from the start. Wallace was always an unlucky man.

On the 10th June we left [Manaus], commencing our voyage very unfortunately for me; for, on going on board, after bidding adieu to my friends, I missed my toucan, which had, no doubt, flown overboard, and not being noticed by any one, was drowned.

His choice of a ship was most unlucky, since she was carrying an inflammable cargo of resin. Three weeks out, on 6 August 1852, the ship caught fire.

I went down into the cabin, now suffocatingly hot and full of smoke, to see what was worth saving. I got my watch and a small tin box containing some shirts and a couple of old note-books, with some drawings of plants and animals, and scrambled up with them on deck. Many clothes and a large portfolio of drawings and sketches remained in my berth; but I did not care to venture down again, and in fact felt a kind of apathy about saving anything, that I can now hardly account for.

The captain at length ordered all into the boats, and was himself the last to leave the vessel.

With what pleasure had I looked upon every rare and curious insect I had added to my collection! How many times, when almost overcome by the ague, had I crawled into the forest and been rewarded by some unknown and beautiful species! How many places, which no European foot but my own had

> trodden, would have been recalled to my memory by the rare birds and insects they had furnished to my collection!
>
> And now everything was gone, and I had not one specimen to illustrate the unknown lands I had trod or to call back the recollection of the wild scenes I had beheld! But such regrets I knew were vain, and I tried to think as little as possible about what might have been and to occupy myself with the state of things which actually existed.

Alfred Wallace returned from the tropics, as Darwin had done, convinced that related species diverge from a common stock, and nonplussed as to why they diverged. What Wallace did not know was that Darwin had hit on the explanation two years after he returned to England from his voyage in the *Beagle*. Darwin recounts that in 1838 he was reading the *Essay on Population* by the Reverend Thomas Malthus ('for amusement', says Darwin, meaning that it was not part of his serious reading) and he was struck by a thought in Malthus. Malthus had said that population multiplies faster than food. If that is true of animals, then they must compete to survive: so that nature acts as a selective force, killing off the weak, and forming new species from the survivors who are fitted to their environment.

'Here then I had at last got a theory by which to work,' says Darwin. And you would think that a man who said that would set to work, write papers, go out and lecture. Nothing of the kind. For four years Darwin did not even commit the theory to paper. Only in 1842 he wrote a draft of thirty-five pages, in pencil; and two years later expanded it to two hundred and thirty pages, in ink. And that draft he deposited with a sum of money and instructions to his wife to publish it if he died.

'I have just finished my sketch of my species theory,' he wrote in a formal letter for her dated 5 July 1844 at Downe, and went on:

> I therefore write this in case of my sudden death, as my most solemn and last request, which I am sure you will consider the same as if legally entered in my Will, that you will devote £400 to its publication, and further, will yourself, or through Hensleigh (Wedgwood), take trouble in promoting it. I wish that my sketch be given to some competent person, with this sum to induce him to take trouble in its improvement and enlargement.
>
> With respect to editors, Mr (Charles) Lyell would be the best if he would undertake it; I believe he would find the work pleasant, and he would learn some facts new to him.
>
> Dr (Joseph Dalton) Hooker would be very good.

We feel that Darwin would really have liked to die before he published the theory, provided after his death the priority should come to him. That is a strange character. It speaks for a man who knew that he was saying something deeply shocking to the public (certainly deeply shocking to his wife) and who was himself, to some extent, shocked by it. The hypochondria (yes, he had some infection from the tropics to excuse it), the bottles of medicine, the enclosed, somewhat suffocating atmosphere of his house and study, the afternoon naps, the delay in writing, the refusal to argue in public: all those speak for a mind that did not want to face the public.

The younger Wallace, of course, was held back by none of these inhibitions. Brashly he went off in spite of all adversities to the Far East in 1854, and for the next eight years travelled all over the Malay archipelago to collect specimens of the wild life there that he would sell in England. By now he was convinced that species are not immutable; he published an essay *On the Law which has regulated the Introduction of New Species* in 1855; and from then 'the question of how changes of species could have been brought about was rarely out of my mind'.

In February of 1858 Wallace was ill on the small volcanic island of Ternate in the Moluccas, the Spice Islands, between New Guinea and Borneo. He had an intermittent fever, was hot and cold by turns, and thought fitfully. And there, on a night of fever, he recalled the same book by Malthus and had the same explanation flash on him that had struck Darwin earlier.

> It occurred to me to ask the question, Why do some die and some live? And the answer was clearly, that on the whole the best fitted lived. From the effects of disease the most healthy escaped; from enemies, the strongest, the swiftest, or the most cunning; from famine, the best hunters or those with the best digestion; and so on.
>
> Then I at once saw, that the ever present variability of all living things would furnish the material from which, by the mere weeding out of those less adapted to the actual conditions, the fittest alone would continue the race.
>
> There suddenly flashed upon me the *idea* of the survival of the fittest.
>
> The more I thought over it, the more I became convinced that I had at length found the long-sought-for law of nature that solved the problem of the *Origin of Species* ... I waited anxiously for the termination of my fit so that I might at once make notes for a paper on the subject. The same evening I did this pretty fully, and on the two succeeding evenings wrote it out carefully in order to send it to Darwin by the next post, which would leave in a day or two.

Wallace knew that Charles Darwin was interested in the subject, and he suggested that Darwin show the paper to Lyell if he thought it made sense.

Darwin received Wallace's paper in his study at Down House four months later, on 18 June 1858. He was at a loss to know what to do. For twenty careful, silent years he had marshalled facts to support the theory, and now there fell on his desk from nowhere a paper of which he wrote laconically on the same day,

> I never saw a more striking coincidence; if Wallace had my MS. sketch written out in 1842, he could not have made a better short abstract!

But friends resolved Darwin's dilemma. Lyell and Hooker, who by now had seen some of his work, arranged that Wallace's paper and one by Darwin should be read in the absence of both at the next meeting of the Linnean Society in London the following month.

The papers made no stir at all. But Darwin's hand had been forced. Wallace was, as Darwin described him, 'generous and noble'. And so Darwin wrote *The Origin of Species* and published it at the end of 1859, and it was instantly a sensation and a best-seller.

The theory of evolution by natural selection was certainly the most important single scientific innovation in the nineteenth century. When all the foolish wind and wit that it raised had blown away, the living world was different because it was seen to be a world in movement. The creation is not static but changes in time in a way that physical processes do not. The physical world ten million years ago was the same as it is today, and its laws were the same. But the living world is not the same; for example, ten million years ago there were no human beings to discuss it. Unlike physics, every generalisation about biology is a slice in time; and it is evolution which is the real creator of originality and novelty in the universe.

If that is so, then each one of us traces his make-up back through the evolutionary process right to the beginnings of life. Darwin, of course, and Wallace looked at behaviour, they looked at bones as they are now, at fossils as they were, to map points on the path by which you and I have come. But behaviour, bones, fossils are already complex systems in life, put together from units which are simpler and must be older. What could the simplest first units be? Presumably they are chemical molecules that characterise life.

So when we look back for the common origin of life, today we look even more deeply, at the chemistry that we all share. The blood in my finger at this moment has come by some millions of steps from the very first primeval molecules that were able to reproduce themselves, over three thousand million years ago. That is evolution in its contemporary conception. The processes by which this has happened in part depend on heredity (which neither Darwin nor Wallace really understood) and in part on chemical structure (which, again, was the province of French scientists rather than British naturalists). The explanations flow together from several fields, but one thing they all have in common. They picture the species separating one after another, in successive stages – that is implied when the theory of evolution is accepted. And from that moment it was no longer possible to believe that life could be re-created at any time now.

When the theory of evolution implied that some animal species came into being more recently than others, critics most often replied by quoting the Bible. Yet most people believed that creation had not stopped with the Bible. They thought that the sun breeds crocodiles from the mud of the Nile. Mice were supposed to grow of themselves in heaps of dirty old clothes; and it was obvious that the origin of bluebottles is bad meat. Maggots must be created inside apples – how else did they get there? All these creatures were supposed to come to life spontaneously, without the benefit of parents.

Fables about creatures that come to life spontaneously are very ancient and are still believed, although Louis Pasteur disproved them beautifully in the 1860s. He did much of that work in his boyhood home in Arbois in the French Jura which he loved to come back to every year. He had done work on fermentation before that, particularly the fermentation of milk (the word 'pasteurisation' reminds us of that). But he was at the height of his power in 1863 (he was forty) when the Emperor of France asked him to look into what goes wrong with the fermentation of wine, and he solved that problem in two years. It is ironic to remember that they were among the best wine years that have ever been; to this day 1864 is remembered as being like no other year.

'The wine is a sea of organisms,' said Pasteur. 'By some it lives, by some it decays.' There are two things striking in that thought. One is that Pasteur found organisms that live without oxygen. At the time that was just a nuisance to wine-growers; but since then it has turned out to be crucial to the understanding of the beginning of life, because then the earth was without oxygen. And second, Pasteur had a remarkable technique by which he could see the traces of life in the liquid. In his twenties he had made his reputation by showing that there are molecules that have a characteristic shape. And he had since shown that this is the thumbprint of their having been through the process of life. That has turned out to be so profound a discovery, and it is still so puzzling, that it is right to look at it in Pasteur's own laboratory and his own words.

> How does one account for the working of the vintage in the vat: of dough left to rise: or the souring of curdling milk: of dead leaves and plants buried in the soil and turning to humus? I must in fact confess that my research has long been dominated by the idea that the structure of substances from the

> point of view of left-handed and right-handedness (if all else is equal) plays an important part in the most intimate laws of the organisation of living beings, and enters into the most obscure corners of their physiology.

Right hand, left hand; that was the deep clue that Pasteur followed in his study of life. The world is full of things whose right-hand version is different from the left-hand version: a right-handed corkscrew as against a left-handed, a right snail as against a left one. Above all, the two hands; they can be mirrored one in the other, but they cannot be turned in such a way that the right hand and the left hand become interchangeable. That was known in Pasteur's time to be true also of some crystals, whose facets are so arranged that there are right-hand versions and left-hand versions.

Pasteur made wooden models of such crystals (he was adroit with his hands, and a beautiful draughtsman) but much more than that, he made intellectual models. In his first piece of research he had hit on the notion that there must be right-handed and left-handed molecules too; and what is true of the crystal must reflect a property of the molecule itself. And that must be displayed by the behaviour of the molecules in any unsymmetrical situation. For instance, when you put them into solution and shine a polarised (that is an unsymmetrical) beam of light through them, the molecules of one kind (say, by convention, the molecules Pasteur called right-handed) must rotate the plane of polarisation of the light to the left. A solution of crystals all of one shape will behave unsymmetrically towards the unsymmetrical beam of light produced in a polarimeter. As the polarising disc is turned, the solution will look alternately dark and light and dark and light again.

The remarkable fact is that a chemical solution from living cells does just that. We still do not know why life has this strange chemical property. But the property establishes that life has a specific chemical

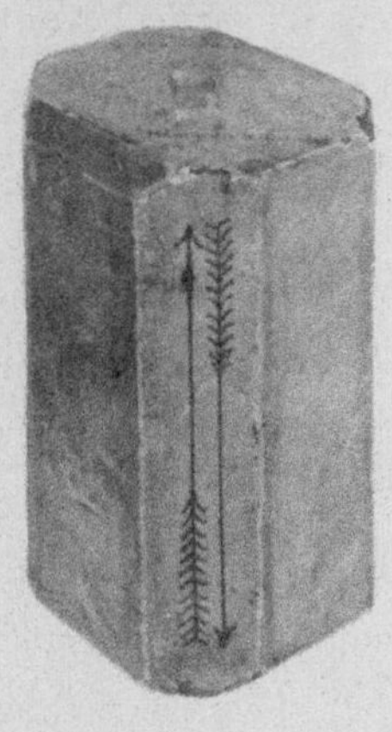

Right hand, left hand; that was the deep clue that Pasteur followed in his study of life. *Pasteur's wooden models of right-handed and left-handed tartrate crystals.*

character, which has maintained itself throughout its evolution. For the first time Pasteur had linked all the forms of life with one kind of chemical structure. From that powerful thought it follows that we must be able to link evolution with chemistry.

The theory of evolution is no longer a battleground. That is because the evidence for it is so much richer and more varied now than it was in the days of Darwin and Wallace. The most interesting and modern evidence comes from our body chemistry. Let me take a practical example: I am able to move my hand at this moment because the muscles contain a store of oxygen, and that has been put there by a protein called myoglobin. That protein is made up of just over one hundred and fifty amino acids. The number is the same in me and all the other animals that use myoglobin. But the amino acids themselves are slightly different. Between me and the chimpanzee there is just one difference in an amino acid; between me and the bush baby (which is a lower primate) there are several amino acid differences; and then between me and the sheep or the mouse, the number of differences increases.

It is the number of amino acid differences which is a measure of the evolutionary distance between me and the other mammals.

It is clear that we have to look for the evolutionary progress of life in a build-up of chemical molecules. And that build-up must begin from the materials that boiled on the earth at its birth. To talk sensibly about the beginning of life we have to be very realistic. We have to ask a historical question. Four thousand million years ago, before life began, when the earth was very young, what was the surface of the earth, what was its atmosphere like?

Very well, we know a rough answer. The atmosphere was expelled from the interior of the earth, and was therefore somewhat like a volcanic neighbourhood anywhere – a cauldron of steam, nitrogen, methane, ammonia and other reducing gases, as well as some carbon dioxide. One gas was absent: there was no free oxygen. That is crucial, because oxygen is produced by the plants and did not exist in a free state before life existed.

These gases and their products, dissolved weakly in the oceans, formed a reducing atmosphere. How would they react next under the action of lightning, electric discharges, and particularly under the action of ultra-violet light – which is very important in every theory of life, because it can penetrate in the absence of oxygen? That question was answered in a beautiful experiment by Stanley Miller in America round about 1950. He put the atmosphere in a flask – the methane, the ammonia, the water, and so on – and went on, for day after day, and boiled and bubbled them up, put an electric discharge through them to simulate lightning and other violent forces. And visibly the mixture darkened. Why? Because on testing it was found that amino acids had been formed in it. That is a crucial step forward, since amino acids are the building blocks of life. From them the proteins are made, and proteins are the constituents of all living things.

We used to think, until a few years ago, that life had to begin in those sultry, electric conditions. And then it began to occur to a few scientists that there is another set of extreme conditions which may be as powerful: that is the presence of ice. It is a strange thought; but ice has two properties which make it very attractive in the formation of simple, basic molecules. First of all, the process of freezing concentrates the material, which at the beginning of time must have been very dilute in the oceans. And secondly, it may be that the crystalline structure of ice makes it possible for molecules to line up in a way which is certainly important at every stage of life.

At any rate, Leslie Orgel did a number of elegant experiments of which I will describe the simplest. He took some of the basic constituents which are sure to have been present in the atmosphere of the earth at any early time: hydrogen cyanide is one, ammonia is another. He made a dilute solution of them in water, and then froze the solution over a period of several days. As a result, the concentrated material is pushed into a sort of tiny iceberg to the top, and there the presence of a small amount of colour reveals that organic molecules have been formed. Some amino acids, no doubt; but, most important, Orgel found that he had formed one of the four fundamental constituents in the genetic alphabet which directs all life. He had made adenine, one of the four bases in DNA. It may indeed be that the alphabet of life in DNA was formed in these sorts of conditions, and not in tropical conditions.

The problem of the origin of life centres not on the complex but on the simplest molecules that will reproduce themselves. It is of the same molecule that characterises life; and the question of the origin of life is therefore the question, whether the basic molecules that have been identified by the work of the present generation of biologists could have been formed by natural processes. We know what we are looking for at the beginning of life: simple, basic molecules like the so-called bases (adenine, thymine, guanine, cytosine) that compose the DNA spirals which reproduce themselves during the division of any cell. The subsequent course by which organisms have become more and more complex is then a different, statistical problem: namely, the evolution of complexity by statistical processes.

It is natural to ask whether self-copying molecules were made many times and in many places. There is no answer to this question except by inferences, which have to be based on our interpretation of the evidence provided by living things today. Life today is controlled by a very few molecules – namely the four bases in DNA. They spell out the message for inheritance in every creature that we know, from a bacterium to an elephant, from a virus to a rose. One conclusion that could be drawn from this uniformity in the alphabet of life is, that these are the only atomic arrangements that will carry out the sequence of replication of themselves.

However, there are not many biologists who believe that. Most biologists think that nature can invent other self-copying arrangements; the possibilities must surely be more numerous than the four we have. If that is right, then the reason why life as we know it is directed by the same four bases is because life *happened* to begin with them. On that interpretation, the bases are evidence that life only began once. After that, when any new arrangement came up, it simply could not link to the living forms that already existed. Certainly no one thinks now that life is still being created from nothing here on earth.

Biology has been fortunate in discovering within the span of one hundred years two great and seminal ideas. One was Darwin's and Wallace's theory of evolution by natural selection. The other was the discovery, by our own contemporaries, of how to express the cycles of life in a chemical form that links them with nature as a whole.

Were the chemicals here on earth at the time when life began unique to us? We used to think so. But the most recent evidence is different. Within the last years there have been found in the interstellar spaces the spectral traces of molecules which we never thought could be formed out in those frigid regions: hydrogen cyanide, cyano acetylene, formaldehyde. These are molecules which we had not supposed to exist elsewhere than on earth. It may turn out that life had more varied beginnings and has more varied forms. And it does not at all follow that the evolutionary path which life (if we discover it) took elsewhere must resemble ours. It does not even follow that we shall recognise it as life – or that it will recognise us.

CHAPTER TEN

WORLD WITHIN WORLD

There are seven basic shapes of crystals in nature, and a multitude of colours. The shapes have always fascinated men, as figures in space and as descriptions of matter; the Greeks thought their elements were actually shaped like the regular solids. And it is true in modern terms that the crystals in nature express something about the atoms that compose them: they help to put the atoms into families. This is the world of physics in our own century, and crystals are a first opening into that world.

Of all the variety of crystals, the most modest is the simple colourless cube of common salt; and yet it is surely one of the most important. Salt has been mined at the great salt mine at Wieliczka near the ancient Polish capital of Cracow for nearly a thousand years, and some of the wooden workings and horse-drawn machinery have been preserved from the seventeenth century. The alchemist Paracelsus may have come this way on his eastern travels. He changed the course of alchemy after AD 1500 by insisting that among the elements that constitute man and nature must be counted salt. Salt is essential to life, and it has always had a symbolic quality in all cultures. Like the Roman soldiers, we still say 'salary' for what we pay a man, though it means 'salt money'. In the Middle East a bargain is still sealed with salt in what the Old Testament calls 'a covenant of salt forever'.

In one respect Paracelsus was wrong; salt is not an element in the modern sense. Salt is a compound of two elements: sodium and chlorine. That is remarkable enough, that a white fizzy metal like sodium, and a yellowish poisonous gas like chlorine, should finish up by making a stable structure, common salt. But more remarkable is that sodium and chlorine belong to families. There is an orderly gradation of similar properties within each family: sodium belongs to the family of alkali metals, and chlorine to the active halogens. The crystals remain unchanged, square and transparent, as we change one member of a family for another. For instance, sodium can certainly be replaced by potassium: potassium chloride. Similarly in the other family the chlorine can be replaced by its sister element bromine: sodium bromide. And, of course, we can make a double change: lithium fluoride, in which sodium has been replaced by lithium, chlorine by fluorine. And yet all the crystals are indistinguishable by the eye.

What makes these family likenesses among the elements? In the 1860s everyone was scratching their heads about that, and several scientists moved towards rather similar answers. The man who solved the problem most triumphantly was a young Russian called Dmitri Ivanovich Mendeleev, who visited the salt mine at Wieliczka in 1859. He was twenty-five then, a poor, modest, hardworking and brilliant young man. The youngest of a vast family of at least fourteen children, he had been the darling of his widowed mother, who drove him through science by her ambition for him.

What distinguished Mendeleev was not only genius, but a passion for the elements. They became his personal friends; he knew every quirk and detail of their behaviour. The elements, of course, were distinguished each by only one basic property, that which John Dalton had proposed originally in 1805: each element has a characteristic atomic weight. How do the properties that make them alike or different flow from that single given constant or parameter? This was the underlying problem and

What distinguished Mendeleev was not only genius but a passion for the elements.

Dmitri Ivanovich Mendeleev.

Mendeleev worked at this. He wrote the elements out on cards, and he shuffled the cards in a game that his friends used to call *Patience*.

Mendeleev wrote on his cards the atoms with their atomic weights, and dealt them out in vertical columns in the order of their atomic weights. The lightest, hydrogen, he did not really know what to do with and he sensibly left it outside his scheme. The next in atomic weight is helium, but luckily Mendeleev did not know that because it had not yet been found on earth – it would have been an awkward maverick until its sister elements were found much later.

Mendeleev therefore began his first column with the element lithium, one of the alkali metals. So it is lithium (the lightest that he knew after hydrogen), then beryllium, then boron, then the familiar elements, carbon, nitrogen, oxygen, and then as the seventh in his column, fluorine. The next element in order of atomic weights is sodium, and since that has a family likeness to lithium, Mendeleev decided this was the place to start again and form a second column parallel to the first. The second column goes on with a sequence of familiar elements: magnesium, aluminium, silicon, phosphorus, sulphur, and chlorine. And sure enough, they make a complete column of seven, so that the last element, chlorine, stands in the same horizontal row as fluorine.

Evidently there is something in the sequence of atomic weights that is not accidental but systematic. It is clear again as we begin the next column, the third. The next elements in order of atomic weights after chlorine are potassium, then calcium. Thus the first row so far contains lithium, sodium, and potassium, which are all alkali metals; and the second row so far contains beryllium, magnesium, and calcium, which are metals with another set of family likenesses. The fact is that the horizontal rows on this arrangement make sense: they put families together. Mendeleev had found, or at least had found evidence for, a mathematical key among the elements. If we arrange them in order of atomic weight, take seven steps to make a vertical column, and start afresh after that with the next column, then we get family arrangements falling together in the horizontal rows.

So far we can follow Mendeleev's scheme without a hitch, just as he set it out in 1871, two years after the first conception. Nothing

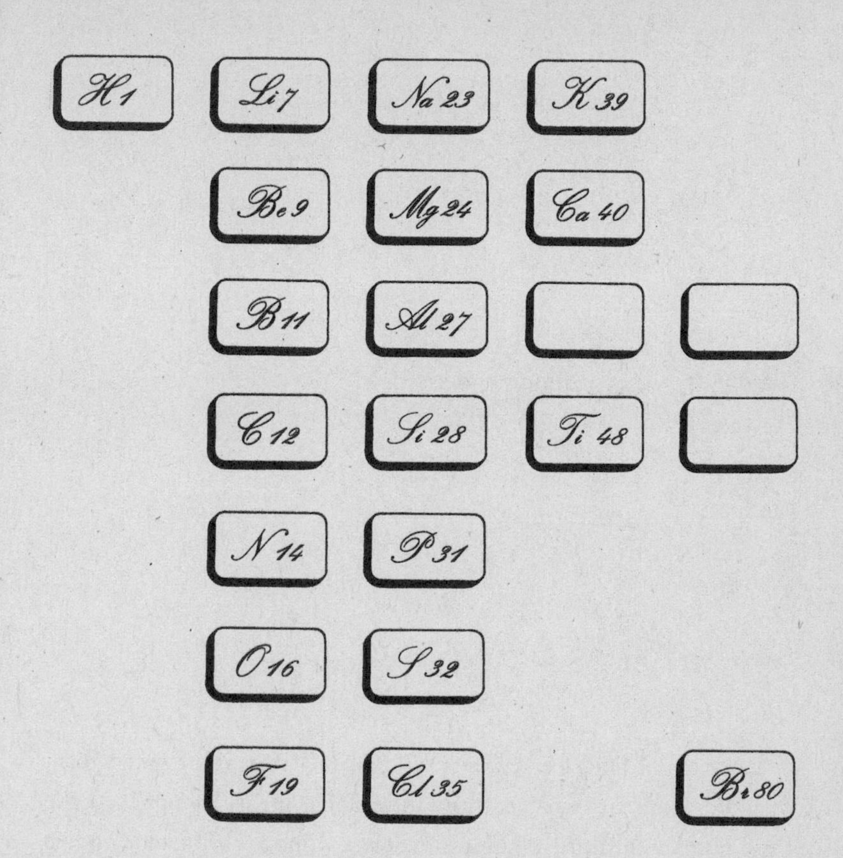

Mendeleev's Game of Patience. The cards are arranged in order of atomic weight: the elements group themselves in families.

falls out of step until the third column – and then, inevitably, the first problem. Why inevitably? Because, as you can see from the case of helium, Mendeleev did not have all the elements. Sixty-three out of the total of ninety-two were known; so sooner or later he was bound to come to gaps. And the first gap he came to was where I stopped, at the third place in the third column.

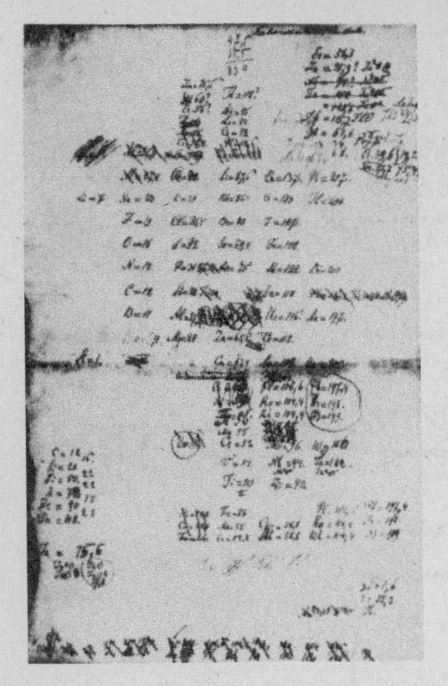

The sequence of atomic weights is not accidental but systematic. *An early draft of Mendeleev's Periodic Table of the Elements of 1869.*

I say that Mendeleev came to a gap, but that abbreviated form of words conceals what is most formidable in his thought. At the third place in the third column Mendeleev came to a difficulty, and he solved the difficulty by *interpreting* it as a gap. He made that choice because the next known element, namely titanium, simply does not have the properties that would fit it there, in the same horizontal row or family with boron and aluminium. So he said, 'There is a missing element there, and when it is found its atomic weight will put it before titanium. Opening the gap will put the later elements of the column into the right horizontal rows; titanium belongs with carbon and silicon' – and indeed it does in the basic scheme.

The conception of the gaps or missing elements was a scientific inspiration. It expressed in practical terms what Francis Bacon had proposed in general terms long ago, the belief that new instances of a law of nature can be guessed or induced in advance from old instances. And Mendeleev's guesses showed that induction is a more subtle process in the hands of a scientist than Bacon and other philosophers supposed. In science we do not simply march along a linear progression of known instances to unknown ones. Rather, we work as in a crossword puzzle, scanning two separate progressions for the points at which they intersect: that is where the unknown instances should be in hiding. Mendeleev scanned the progression of atomic weights in the columns, and the family likenesses in the rows, to pinpoint the missing elements at their intersections. By doing so, he made practical predictions, and he also made manifest (what is still poorly understood) how scientists actually carry out the process of induction.

Very well: the points of greatest interest are the gaps that lie in the third and fourth columns. I will not go on building the table beyond there – except to say that when you count the gaps and go on down, sure enough, the column ends where it should, at bromine in the halogen family. There were a number of gaps, and Mendeleev singled out three. The first I have just pointed to in the third column and third row. The other two are in the fourth column, in the third and fourth rows. And of them Mendeleev prophesied that on discovery it would be found, not only that they have atomic weights that fit into the vertical progression, but that they would have those properties that are appropriate to the families in the third and fourth horizontal rows.

For instance, the most famous of Mendeleev's forecasts, and the last to be confirmed, was the third – what he called eka-silicon. He predicted the properties of this strange and important element with great exactitude, but it was nearly twenty years before it was found in Germany, and called not after Mendeleev, but *germanium*. Having begun from the principle that 'eka-silicon will have properties intermediate between silicon and tin', he had predicted that it would be 5.5 times heavier than water; that was right. He predicted that its oxide would be 4.7 times heavier than water; that was right. And so on with chemical and other properties.

These forecasts made Mendeleev famous everywhere – except in Russia: he was not a prophet there, because the Tsar did not like his liberal politics. The later discovery in England of a whole new row of elements, beginning with helium, neon, argon, enlarged his triumph. He was not elected to the Russian Academy of Sciences, but in the rest of the world his name was magic.

The underlying pattern of the atoms is numerical, that was clear. And yet that cannot be the whole story; we must be missing something. It simply does not make sense to believe that all the properties of the elements are contained in one number, the atomic weight: which hides – what? The weight of an atom might be a measure of its complexity. If so, it must hide some internal structure, some way the atom is physically put together, which generates those properties. But, of course, as an idea that was inconceivable so long as it was believed that the atom is indivisible.

And that is why the turning-point comes in 1897, when J. J. Thomson in Cambridge discovers the electron. Yes, the atom has constituent parts; it is not indivisible, as its Greek name had implied. The electron is a tiny part of its mass or weight, but a real part, and it carries a single electric charge. Each element is characterised by the number of electrons in its atoms. And their number is exactly equal to the number of the place in Mendeleev's table that that element occupies when hydrogen and helium are included in first and second place. That is, lithium has three electrons, beryllium has four electrons, boron has five, and so on steadily all through the table. The place in the table that an element occupies is called its atomic number, and now that turned out to stand for a physical reality within its atom – the number of electrons there. The picture has shifted from atomic weight to atomic number, and that means, essentially, to atomic structure.

That is the intellectual breakthrough with which modern physics begins. Here the great age opens. Physics becomes in those years the greatest collective work of science – no, more than that, the great collective work of art of the twentieth century.

I say 'work of art', because the notion that there is an underlying structure, a world within the world of the atom, captured the imagination of artists at once. Art from the year 1900 on is different from the art before it, as can be seen in any original painter of the time: Umberto Boccioni, for instance, in *The Forces of a Street*, or his *Dynamism of a Cyclist.* Modern art begins at the same time as modern physics because it begins in the same ideas.

Since the time of Newton's *Opticks*, painters had been entranced by the coloured surface of things. The twentieth century changed that. Like the X-ray pictures of Röntgen, it looked for the bone beneath the skin, and for the deeper, solid structure that builds up from inside the total form of an object or a body. A painter like Juan Gris is engaged in the analysis of structure, whether he is looking at natural forms in *Still Life* or at the human form in *Pierrot.*

The Cubist painters, for example, are obviously inspired by the families of crystals. They see in them the shape of a village on a hillside, as Georges Braque did in his *Houses at L'Estaque*, or a group of women as Picasso painted them in *Les Demoiselles d'Avignon.* In Pablo Picasso's famous beginning to Cubist painting – a single face, the *Portrait of Daniel-Henry Kohnweiler* – the interest has shifted from the skin and the features to the underlying geometry. The head has been taken apart into mathematical shapes and then put together as a reconstruction, a re-creation, from the inside out.

This new search for the hidden structure is striking in the painters of Northern Europe: Franz Marc, for example, looking at the natural landscape in *Deer in a Forest*; and (a favourite with scientists) the Cubist Jean Metzinger, whose *Woman on a Horse* was owned by Niels Bohr, who collected pictures in his house in Copenhagen.

There are two clear differences between a work of art and a scientific paper. One is that in the work of art the painter is visibly taking the world to pieces and putting it together on the same canvas. And the other is that you can watch him thinking while he is doing it. (For example, Georges Seurat putting one coloured dot beside another of a different colour to get the total effect in *Young Woman with a Powder Puff* and *Le Bec*.) In both those respects the scientific paper is often deficient. It often is only analytic; and it almost always hides the process of thought in its impersonal language.

I have chosen to talk about one of the founder fathers of twentieth-century physics, Niels Bohr, because in both these respects he was a consummate artist. He had no ready-made answers. He used to begin his lecture courses by saying to his students, 'Every sentence that I utter should be regarded by you not as an assertion but as a question'. What he questioned was the structure of the world. And the people that he worked with, when young and old (he was still penetrating in his seventies), were others who were taking the world to pieces, thinking it out, and putting it together.

He went first in his twenties to work with J. J. Thomson, and his one-time student Ernest Rutherford who, round about 1910, was the outstanding experimental physicist in the world. (Thomson and Rutherford had both been turned to science by the interest of their widowed mothers, as Mendeleev had been.) Rutherford was then a professor at Manchester University. And in 1911 he had proposed a new model for the atom. He had said that the bulk of the atom is in a heavy nucleus or core at the centre, and the electrons circle it on orbiting paths, the way that the planets circle the sun. It was a brilliant conception – and a nice irony of history, that in three hundred years the outrageous image of Copernicus and Galileo and Newton had become the most natural model for every scientist. As often in science, the incredible theory of one age had become the everyday image for its successors.

Nevertheless, there was something wrong with Rutherford's model. If the atom is really a little machine, how can its structure account for the fact that it does not run down – that it is a little perpetual motion machine, and the only perpetual motion machine that we have? The planets as they move in their orbits lose energy continuously, so that year by year their orbits get smaller – a very little smaller, but in time they will fall into the sun. If the electrons are exactly like the planets, then they will fall into the nucleus. There must be something to stop the electrons from losing energy continuously. That required a new principle in physics, so as to limit the energy an electron can give out to fixed values. Only so can there be a yardstick, a definite unit which holds the electrons to orbits of fixed sizes.

Niels Bohr discovered the unit he was looking for in the work that Max Planck had published in Germany in 1900. What Planck had shown, a dozen years earlier, is that in a world in which matter comes in lumps, energy must come in lumps, or quanta, also. By hindsight that does not seem so strange. But Planck knew how revolutionary the idea was the day he had it, because on that day he took his little boy for one of those professorial walks that academics take after lunch all over the world, and said to him, 'I have had a conception today as revolutionary and as great as the kind of thought that Newton had'. And so it was.

Now in a sense, of course, Bohr's task was easy. He had the Rutherford atom in one hand, he had the quantum in the other. What was there so wonderful about a young man of twenty-seven in 1913 putting the two together and making the modern image of the atom? Nothing but the wonderful, visible thought-process: nothing but the effort of synthesis. And the idea of seeking support for it in the one place where it could be found: the fingerprint of the atom, namely the spectrum in which its behaviour becomes visible to us, looking at it from outside.

That was Bohr's marvellous idea. The inside of the atom is invisible, but there is a window in it, a stained-glass window: the spectrum of the atom. Each element has its own spectrum, which is not continuous like that which Newton got from white light, but has a number of bright lines which characterise that element. For example, hydrogen has three rather vivid lines in its visible spectrum: a red line, a blue-green line, and a blue line. Bohr explained them each as a release of energy when the single electron in the hydrogen atom jumps from one of the outer orbits to one of the inner orbits.

As long as the electron in a hydrogen atom remains in one orbit, it emits no energy. Whenever it jumps from an outer orbit to an inner orbit, the energy difference between the two is emitted as a light quantum. These emissions from many billions of atoms simultaneously are what we see as a characteristic hydrogen line. The red line is when the electron jumps from the third orbit to the second; the blue-green line when the electron jumps from the fourth orbit to the second.

Bohr's paper *On the Constitution of Atoms and Molecules* became a classic at once. The structure of the atom was now as mathematical as Newton's universe. But it contained the additional principle of the quantum. Niels Bohr had built a world inside the atom by going beyond the laws of physics as they had stood for two centuries after Newton. He returned to Copenhagen in triumph. Denmark was home for him again, a new place to work. In 1920 they built the Niels Bohr Institute in Copenhagen for him. Young men came there to discuss quantum physics from Europe, America, and the Far East.

Werner Heisenberg came often from Germany and was goaded into conceiving some of his crucial ideas there: Bohr would never allow anyone to stop at a half-formed idea.

It is interesting to trace the steps of confirmation of Bohr's model of the atom, because in a way they recapitulate the life-cycle of every scientific theory. First comes the paper. In that, known results are used to support the model. That is to say, the spectrum of hydrogen in particular is shown to have lines, long known, whose positions correspond to quantum transitions of the electron from one orbit to another.

The next step is to extend that kind of confirmation to a new phenomenon: in this case, lines in the higher energy X-ray spectrum, which is not visible to the eye but which is formed in just the same way by electron leaps. That work was going on in Rutherford's laboratory in 1913, and yielded beautiful results exactly confirming what Bohr had predicted. The man who did the work was Harry Moseley, twenty-seven years old, who did no more brilliant work because he died in the forlorn British attack at Gallipoli in 1915 – a campaign which cost, indirectly, the lives of other young men of high promise, among them that of the poet Rupert Brooke. Moseley's work, like Mendeleev's, suggested some missing elements, and one of them was discovered in Bohr's laboratory and named *hafnium*, after the Latin name for Copenhagen. Bohr announced the discovery incidentally in the speech he made when accepting the Nobel Prize for Physics in 1922. The theme of the speech is memorable, for Bohr described in detail what he summarised almost poetically in another speech: how the concept of the quantum had

> led gradually to a systematic classification of the types of stationary binding of any electron in an atom, offering a

> complete explanation of the remarkable relationships between the physical and chemical properties of the elements, as expressed in the famous periodic table of Mendeleev. Such an interpretation of the properties of matter appeared as a realisation, even surpassing the dreams of the Pythagoreans, of the ancient ideal of reducing the formulation of the laws of nature to considerations of pure numbers.

And just at this moment, when everything seems to be going so swimmingly, we suddenly begin to realise that Bohr's theory, like every theory sooner or later, is reaching the limits of what it can do. It begins to develop little cranky weaknesses, a kind of rheumatic pain. And then comes the crucial realisation that we have not cracked the real problem of atomic structure at all. We have cracked the shell. But within that shell the atom is an egg with a yolk, the nucleus; and we have not begun to understand the nucleus.

Niels Bohr was a man with a taste for contemplation and leisure. When he won the Nobel Prize he spent the money on buying a house in the country. His taste for the arts also ran to poetry. He said to Heisenberg, 'When it comes to atoms, language can be used only as in poetry. The poet, too, is not nearly so concerned with describing facts as with creating images'. That is an unexpected thought: when it comes to atoms, language is not describing facts but creating images. But it is so. What lies below the visible world is always imaginary, in the literal sense: a play of images. There is no other way to talk about the invisible – in nature, in art, or in science.

When we step through the gateway of the atom, we are in a world which our senses cannot experience. There is a new architecture there, a way that things are put together which we cannot know: we only try to picture it by analogy, a new act of imagination. The architectural images come from the concrete world of our senses, because that is the only world that words describe. But all our ways of picturing the invisible are metaphors, likenesses that we snatch from the larger world of eye and ear and touch.

Once we have discovered that the atoms are not the ultimate building blocks of matter, we can only try to make models of how the building blocks link and act together. The models are meant to show, by analogy, how matter is built up. So, to test the models, we have to take matter to pieces, like the diamond cleaver feeling for the structure of the crystal.

The ascent of man is a richer and richer synthesis, but each step is an effort of analysis: of deeper analysis, world within world. When the atom was found to be divisible it seemed that it might have an indivisible centre, the nucleus. And then it turned out, around 1930, that the model needed a new refinement. The nucleus at the centre of the atom is not the ultimate fragment of reality either.

At twilight on the sixth day of Creation, so say the Hebrew commentators to the Old Testament, God made for man a number of tools that give him also the gift of creation. If the commentators were alive today, they would write 'God made the neutron'. Here it is, at Oak Ridge in Tennessee, the blue glow that is the trace of neutrons: the visible finger of God touching Adam in Michelangelo's painting, not with breath but with power.

I must not start quite so early. Let me begin the story about 1930. At that time the nucleus of the atom still seemed as invulnerable as the atom itself had once seemed. The trouble was that there was no way it could come apart into electrical pieces: the numbers simply would not fit. The nucleus has a positive charge (to balance the electrons in the atom) equal to the atomic number. But the mass of the nucleus is not a constant multiple of the charge: it ranges from being equal to the charge (in hydrogen) to much over twice the charge in the heavy elements. That was inexplicable, so long as everyone remained convinced that all matter must be built up from electricity.

It was James Chadwick who broke with that deeply rooted idea, and proved in 1932 that the nucleus consists of two kinds of particles: not only of the electrical positive proton, but of a nonelectrical particle, the neutron. The two particles are almost equal in mass, namely equal (roughly) to the atomic weight of hydrogen. Only the simplest nucleus of hydrogen contains no neutrons, and consists of a single proton.

The neutron was therefore a new kind of probe, a sort of alchemist's flame, because, having no electric charge, it could be fired into the nuclei of atoms without suffering electrical disturbance, and change them. The modern alchemist, the man who more than anyone took advantage of that new tool, was Enrico Fermi in Rome.

Enrico Fermi was a strange creature. I did not know him until much later, because in 1934 Rome was in the hands of Mussolini, Berlin was in the hands of Hitler, and men like me did not travel there. But when I saw him in New York, later, he struck me as the cleverest man I had ever set eyes on – well, perhaps the cleverest man with one exception. He was compact, small, powerful, penetrating, very sporty, and always with the direction in which he was going as clear in his mind as if he could see to the very bottom of things.

Fermi set about shooting neutrons at every element in turn, and the fable of transmutation came true in his hands. The neutrons he used you can see streaming out of a reactor because it is what is lightly called a 'swimming pool' reactor, meaning that the neutrons are slowed down by water. I should give it its proper name: it is a High Flux Isotope Reactor, which has been developed at Oak Ridge, Tennessee.

Transmutation was, of course, an age-old dream. But to men like me, with a theoretical bent of mind, what was most exciting about the 1930s was that there began to open up the evolution of nature. I must explain that phrase. I began here by talking about the day of Creation, and I will do that again. Where shall I start? Archbishop James Ussher of Armagh, a long time ago, about 1650, said that the universe was created in 4004 BC. Armed as he was with dogma and ignorance, he brooked no rebuttal. He or another cleric knew the year, the date, the day of the week, the hour, which fortunately I have forgotten. But the puzzle of the age of the world remained, and remained a paradox, well into the 1900s: because, while it was then clear that the earth was many, many millions of years old, we could not conceive where the energy came from in the sun and the stars to keep them going so long. By then we had Einstein's equations, of course, which showed that the loss of matter would produce energy. But how was the matter rearranged?

Very well: that is really the crux of energy and the door of understanding that Chadwick's discovery opened. In 1939 Hans Bethe, working at Cornell University, for the first time explained in very precise terms the transformation of hydrogen to helium in the sun, by which a loss of mass streams out to us as this proud gift of energy. I speak of these matters with a kind of passion, because of course to me they have the quality, not of memory, but of experience. Hans Bethe's explanation is as vivid to me as my own wedding day, and the subsequent steps that followed as the birth of my own children. Because what was revealed in the years that followed (and finally scaled in what I suppose to be the definitive analysis in 1957) is that in all the stars there are going on processes which build up the atoms one by one into more and more complex structures. Matter itself evolves. The word comes from Darwin and biology, but it is the word that changed physics in my lifetime.

The first step in the evolution of the elements takes place in young stars, such as the sun. It is the step from hydrogen to helium, and it needs the great heat of the interior; what we see on the surface of the sun are only storms produced by that action. (Helium was first identified by a spectrum line during the eclipse of the sun in 1868; that is why it was called helium, for it was not known on earth then.) What happens in effect is that from time to time a pair of nuclei of heavy hydrogen collide and fuse to make a nucleus of helium.

In time the sun will become mostly helium. And then it will become a hotter star in which helium nuclei collide to make heavier atoms in turn. Carbon, for instance, is formed in a star whenever three helium nuclei collide at one spot within less than a millionth of a millionth of a second. Every carbon atom in every living creature has been formed by such a wildly improbable collision. Beyond carbon, oxygen is formed, silicon, sulphur and heavier elements. The most stable elements are in the middle of Mendeleev's table, roughly between iron and silver. But the process of building the elements overshoots well beyond them.

If the elements are built up one by one, why does nature stop? Why do we find only ninety-two elements, of which the last is uranium? To answer that question, we have, evidently, to build elements beyond it, and to confirm that as the elements become bigger, they become more complex and tend to fall apart into pieces. When we do that, however, we are not only making new elements but are making something that is potentially explosive. The element plutonium, which Fermi made in the first historic Graphite Reactor (we called it a 'Pile' in those old colloquial days) was the man-made element that demonstrated this to the world at large. In part it is a monument to the genius of Fermi; but I think of it as a tribute to the god Pluto of the underworld who gave his name to the element, for forty thousand people died at Nagasaki of the plutonium bomb there. It is one more time in the history of the world when a monument commemorates a great man and many dead, together.

The first historic graphite reactor.
Experimental graphite-uranium pile designed by the group under Enrico Fermi, which went into operation for the first time on 2 December 1942 on the squash court, West Stands, Stagg Field, University of Chicago.

I must return briefly to the mine at Wieliczka because there is a historical contradiction to be explained here. The elements are being built up in the stars constantly, and yet we used to think that the universe is running down. Why? Or how? The idea that the universe is running down comes from a simple observation about machines. Every machine consumes more energy than it renders. Some of it is wasted in friction, some of it is wasted in wear. And in some more sophisticated machines than the ancient wooden capstans at Wieliczka, it is wasted in other necessary ways – for example, in a shock-absorber or a radiator. These are all ways in which the energy is degraded. There is a pool of inaccessible energy into which some of the energy that we put in always runs, and from which it cannot be recovered.

In 1850 Rudolf Clausius put that thought into a basic principle. He said that there is energy which is available, and there is also a residue of energy which is not accessible. This inaccessible energy he called entropy, and he formulated the famous Second Law of Thermodynamics: entropy is always increasing. In the universe, heat is draining into a sort of lake of equality in which it is no longer accessible.

That was a nice idea a hundred years ago, because then heat could still be thought of as a fluid. But heat is not material any more than fire is, or any more than life is. Heat is a random motion of the atoms. And it was Ludwig Boltzmann in Austria who brilliantly seized on that idea to give a new interpretation to what happens in a machine, or a steam engine, or the universe.

When energy is degraded, said Boltzmann, it is the atoms that assume a more disorderly state. And entropy is a measure of disorder: that is the profound conception that came from Boltzmann's new interpretation. Strangely enough, a measure of disorder can be made; it is the probability of the particular state – defined here as the number of ways it can be assembled from its atoms. He put that quite precisely,

$$S = K \log W;$$

S, the entropy, is to be represented as proportional to the logarithm of W, the probability of the given state (K being the constant of proportionality which is now called Boltzmann's constant).

Of course, disorderly states are much more probable than orderly states, since almost every assembly of the atoms at random will be disorderly; so by and large any orderly arrangement will run down.

But 'by and large' is not 'always'. It is not true that orderly states *constantly* run down to disorder. It is a statistical law, which means that order will *tend* to vanish. But statistics do not say 'always'. Statistics allow order to be built up in some islands of the universe (here on earth, in you, in me, in the stars, in all sorts of places) while disorder takes over in others.

That is a beautiful conception. But there is still one question to be asked. If it is true that probability has brought us here, is not the probability so low that we have no right to be here?

People who ask that question always picture it thus. Think of all the atoms that make up my body at this moment. How madly improbable that they should come to this place at this instant and form me. Yes, indeed, if that was how it happened, it would not only be improbable – I would be virtually impossible.

But, of course, that is not how nature works. Nature works by steps. The atoms form molecules, the molecules form bases, the bases direct the formation of amino acids, the amino acids form proteins, and proteins work in cells. The cells make up first of all the simple animals, and then sophisticated ones, climbing step by step. The stable units that compose one level or stratum are the raw material for random encounters which produce higher configurations, some of which will chance to be stable. So long as there remains a potential of stability which has not become actual, there is no other way for chance to go. Evolution is the climbing of a ladder from simple to complex by steps, each of which is stable in itself.

Since this is very much my subject, I have a name for it: I call it *Stratified Stability*. That is what has brought life by slow steps but constantly up a ladder of increasing complexity – which is the central progress and problem in evolution. And now we know that that is true not only of life but of matter. If the stars had to build a heavy element like iron, or a super-heavy element like uranium, by the instant assembly of all the parts, it would be virtually impossible. No. A star builds hydrogen to helium; then at another stage in a different star helium is assembled to carbon, to oxygen, to heavy elements; and so step by step up the whole ladder to make the ninety-two elements in nature.

We cannot copy the processes in the stars as a whole, because we do not command the immense temperatures that are needed to fuse most elements. But we have begun to put our foot on the ladder: to copy the first step, from hydrogen to helium. In another part of Oak Ridge the fusion of hydrogen is attempted.

It is hard to recreate the temperature within the sun, of course – over ten million degrees centigrade. And it is still harder to make any kind of container that will survive that temperature and trap it for even a fraction of a second. There are no materials that will do; a container for a gas in this violent state can only have the form of a magnetic trap. This is a new kind of physics: plasma-physics. Its excitement, yes, and its importance, is that it is the physics of nature. For once, the rearrangements that man makes run, not against the direction of nature, but along the same steps which nature herself takes in the sun and in the stars.

Immortality and mortality is the contrast on which I end this essay. Physics in the twentieth century is an immortal work. The human imagination working communally has produced no monuments to equal it, not the pyramids, not the *Iliad*, not the ballads, not the cathedrals. The men who made these conceptions one after another are the pioneering heroes of our age. Mendeleev, shuffling his cards; J. J. Thomson, who overturned the Greek belief that the atom is indivisible; Rutherford, who turned it into a planetary system; and Niels Bohr, who made that model work. Chadwick, who discovered the neutron, and Fermi, who used it to open up and to transform the nucleus. And at the head of them all are the iconoclasts, the first founders of the new conceptions: Max Planck, who gave energy an atomic character like matter; and Ludwig Boltzmann to whom, more than anyone else, we owe the fact that the atom – the world within a world – is as real to us now as our own world.

Who would think that, only in 1900, people were battling, one might say to the death, over the issue whether atoms are real or not. The great philosopher Ernst Mach in Vienna said, No. The great chemist Wilhelm Ostwald said, No. And yet one man, at that critical turn of the century, stood up for the reality of atoms on fundamental grounds of theory. He was Ludwig Boltzmann, at whose memorial I pay homage.

Boltzmann was an irascible, extraordinary, difficult man, an early follower of Darwin, quarrelsome and delightful, and everything that a human being should be. The ascent of man teetered on a fine intellectual balance at that point, because had anti-atomic doctrines then really won the day, our advance would certainly have been set back by decades, and perhaps a hundred years. And not only in physics would it have been held back, but in biology, which is crucially dependent on that. Did Boltzmann just argue? No. He lived and died that passion. In 1906, at the age of sixty-two, feeling isolated and defeated, at the very moment when atomic doctrine was going to win, he thought all was lost, and he committed suicide. What remains to commemorate him is his immortal formula,

$$S = K \log W,$$

carved on his grave.

I have no phrase to match the compact and penetrating beauty of Boltzmann's. But I will take a quotation from the poet William Blake, who begins the *Auguries of Innocence* with four lines:

> To see a World in a Grain of Sand
> And a Heaven in a Wild Flower
> Hold Infinity in the palm of your hand
> And Eternity in an hour.

CHAPTER ELEVEN

KNOWLEDGE OR CERTAINTY

One aim of the physical sciences has been to give an exact picture of the material world. One achievement of physics in the twentieth century has been to prove that that aim is unattainable.

Take a good, concrete object, the human face. I am listening to a blind woman as she runs her fingertips over the face of a man she senses for the first time, thinking aloud. 'I would say that he is elderly. I think, obviously, he is not English. He has a rounder face than most English people. And I should say he is probably Continental, if not Eastern-Continental. The lines in his face would be lines of possible agony. I thought at first they were scars. It is not a happy face.'

This is the face of Stephan Borgrajewicz, who like me was born in Poland. In 'Portrait of Stephan Borgrajewicz' it is seen by the Polish artist, Feliks Topolski. We are aware that these pictures do not so much fix the face as explore it; that the artist is tracing the detail almost as if by touch; and that each line that is added strengthens the picture but never makes it final. We accept that as the method of the artist.

But what physics has now done is to show that that is the only method to knowledge. There is no absolute knowledge. And those who claim it, whether they are scientists or dogmatists, open the door to tragedy. All information is imperfect. We have to treat it with humility. That is the human condition; and that is what quantum physics says. I mean that literally.

Look at the face across the whole spectrum of electromagnetic information. The question I am going to ask is: How fine and how exact is the detail that we can see with the best instruments in the world – even with a perfect instrument, if we can conceive one?

And seeing the detail need not be confined to seeing with visible light. James Clerk Maxwell in 1867 proposed that light is an electromagnetic wave, and the equations that he constructed for it implied that there are others. The spectrum of visible light, from red to violet, is only an octave or so in the range of invisible radiations. There is a whole keyboard of information, all the way from the longest wavelengths of radiowaves (the low notes) to the shortest wavelengths of X-rays and beyond (the highest notes). We will shine it all, turn by turn, on the human face.

The longest of the invisible waves are the radiowaves, whose existence Heinrich Hertz proved nearly a hundred years ago in 1888, and so confirmed Maxwell's theory. Because they are the longest, they are also the crudest. A radar scanner working at a wavelength of a few metres will not see the face at all, unless we make the face also some metres across, like a Mexican stone head. Only when we shorten the wavelength does any detail appear on the giant head: at a fraction of a metre, the ears. And at the practical limit of radiowaves, a few centimetres, we detect the first trace of the man beside the statue.

Next we look at the face, the man's face, with a camera which is sensitive to the next range of radiation, to wavelengths of less than a millimetre, infra-red rays. The astronomer William Herschel discovered them in 1800, by noticing the warmth when he focused his telescope beyond red light; for the infrared rays are heat rays. The camera plate translates them into visible light in a rather arbitrary code, making the hottest look blue and the coolest look red or dark. We see the rough features of the face: the eyes, the mouth, the nose – we see the heat steam from the nostrils. We learn something new about the human face, yes. But what we learn has no detail.

At its shortest wavelengths, some hundredths of a millimetre or less, infra-red shades gently into visible red. The film that we use now is sensitive to both, and the face springs to life. It is no longer a man, it is the man we know: Stephan Borgrajewicz.

White light reveals him to the eye visibly, in detail; the small hairs, the pores in the skin, a blemish here, a broken vessel there. White light is a mixture of wavelengths, from red to orange to yellow to green to blue and finally to violet, the shortest visible waves. We ought to see more exact details with the short violet waves than the long red waves. But in practice, a difference of an octave or so does not help much.

The painter analyses the face, takes the features apart, separates the colours, enlarges the image. It is natural to ask, Should not the scientist use a microscope to isolate and analyse the finer features? Yes, he should. But we ought to understand that the microscope enlarges the image but cannot improve it: the sharpness of detail is fixed by the wavelength of the light. The fact is that at any wavelength we can intercept a ray only by objects about as large as a wavelength itself; a smaller object simply will not cast a shadow.

An enlargement of over two hundred times can single out an individual cell in the skin with ordinary white light. But to get more detail, we need a still shorter wavelength. The next step, then, is ultra-violet light, which has a wavelength of ten thousandth of a millimetre and less – shorter by a factor of ten and more than visible light. If our eyes were able to see into the ultra-violet, they would see a ghostly landscape of fluorescence. The ultra-violet microscope looks through the shimmer into the cell, enlarged three thousand five hundred times, to the level of single chromosomes. But that is the limit: no light will see the human genes within a chromosome.

Once again, to go deeper, we must shorten the wavelength: next, to the X-rays. However, they are so penetrating that they cannot be focused by any material; we cannot build an X-ray microscope. So we must be content to fire them at the face and get a sort of shadow. The detail depends now on their penetration. We see the skull beneath the skin – for example, that the man has lost his teeth. This probing of the body made X-rays exciting as soon as Wilhelm Konrad Röntgen discovered them in 1895, because here was a finding in physics that seemed designed by nature to serve medicine. It made Röntgen a kindly father figure, and he was the hero who won the first Nobel prize in 1901.

A lucky chance in nature will sometimes let us do more by

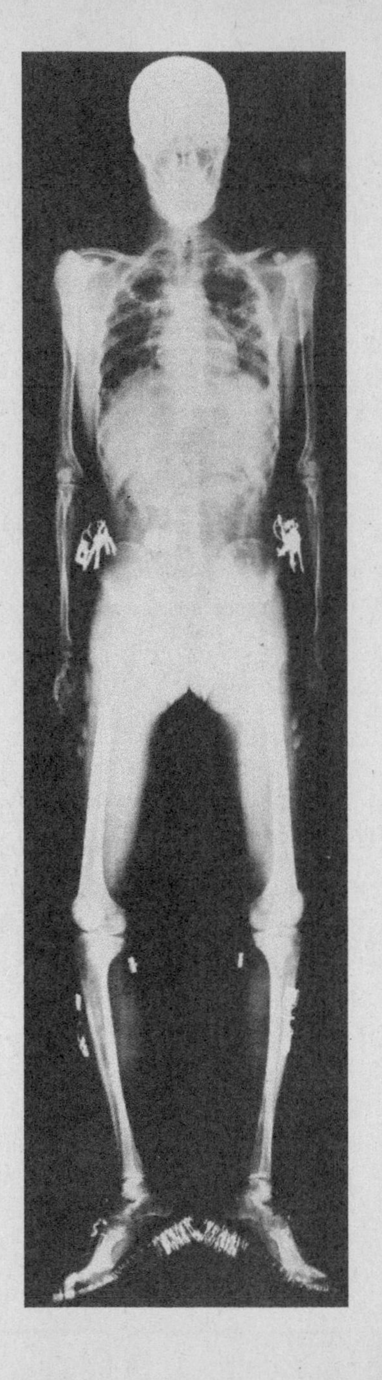

The X-rays form a regular pattern of ripples from which the position of the obstructing atoms can be inferred.
X-ray diffraction pattern of a crystal of DNA.

(Opposite) The probing of the body made X-rays exciting as soon as Röntgen discovered them.
Röntgen's original plate of a man in his shoes, with his keys in his trouser pockets.

a flanking movement, that is, by inferring an arrangement which cannot be seen directly. X-rays will not show us an individual atom, because it is too small to cast a shadow even at this short wavelength. Nevertheless, we can map the atoms in a crystal because their spacing is regular, so that the X-rays will form a regular pattern of ripples from which the positions of the obstructing atoms can be inferred. This is the pattern of atoms in the DNA spiral: this is what a gene is like. The method was invented in 1912 by Max von Laue, and was a double stroke of ingenuity, for it was the first proof that atoms are real, and also the first proof that X-rays are electromagnetic waves.

We have one step more left to take, to the electron microscope, where the rays are so concentrated that we no longer know whether to call them waves or particles. Electrons are fired at an object, and they trace its outline like a knife-thrower at a fair. The smallest object that has ever been seen is a single atom of thorium. It is spectacular. And yet the soft image confirms that, like the knives that graze the girl at the fair, even the hardest electrons do not give a hard outline. The perfect image is still as remote as the distant stars.

We are here face to face with the crucial paradox of knowledge. Year by year we devise more precise instruments with which to observe nature with more fineness. And when we look at the observations, we are discomfited to see that they are still fuzzy, and we feel that they are as uncertain as ever. We seem to be running after a goal which lurches away from us to infinity every time we come within sight of it.

The paradox of knowledge is not confined to the small, atomic scale; on the contrary, it is as cogent on the scale of man, and even of the stars. Let me put it in the context of an astronomical observatory. Karl Friedrich Gauss's observatory at Göttingen was built about 1807. Throughout his lifetime and ever since (the best part of two hundred years) astronomical instruments have been improved. We

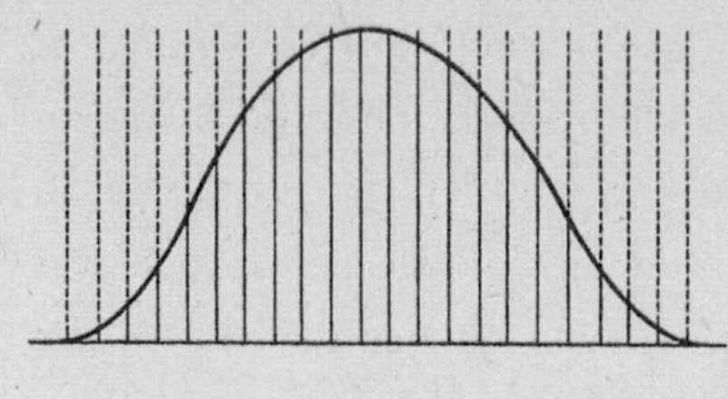

The paradox of knowledge is not confined to the small, atomic scale; on the contrary, it is as cogent on the scale of man, and even the stars. *Karl Friedrich Gauss. The Gaussian curve.*

look at the position of a star as it was determined then and now, and it seems to us that we are closer and closer to finding it precisely. But when we actually compare our individual observations today, we are astonished and chagrined to find them as scattered within themselves as ever. We had hoped that the human errors would disappear, and that we would ourselves have God's view. But it turns out that the errors cannot be taken out of the observations. And that is true of stars, or atoms, or just looking at somebody's picture, or hearing the report of somebody's speech.

Gauss recognised this with that marvellous, boyish genius that he had right up to the age of nearly eighty at which he died. When he was only eighteen years old, when he came to Göttingen to enter the University in 1795, he had already solved the problem of the best estimate of a series of observations which have internal errors. He reasoned then as statistical reasoning still goes today.

When an observer looks at a star, he knows that there is a multitude of causes for error. So he takes several readings, and he hopes, naturally, that the best estimate of the star's position is the average – the centre of the scatter. So far, so obvious. But Gauss pushed on to ask what the *scatter* of the errors tells us. He devised the Gaussian curve in which the *scatter* is summarised by the deviation, or spread, of the curve. And from this came a far-reaching idea: the scatter marks an area of uncertainty. We are not sure that the true position is the centre. All we can say is that it lies *in the area of uncertainty*, and the area is calculable from the observed scatter of the individual observations.

Having this subtle view of human knowledge, Gauss was particularly bitter about philosophers who claimed that they had a road to knowledge more perfect than that of observation. Of many examples I will choose one. It happens that there is a philosopher called Friedrich Hegel, whom I must confess I specifically detest. And I am happy to share that profound feeling with a far greater man, Gauss. In 1800 Hegel presented a thesis, if you please, proving that although the definition of planets had changed since the Ancients, there still could only be, philosophically, seven planets. Well, not only Gauss knew how to answer that: Shakespeare had answered that long before. There is a marvellous passage in *King Lear*, in which who else but the Fool says to the King: 'The reason why the seven Starres are no mo then seuen, is a pretty reason'. And the King wags sagely and says: 'Because they are not eight'. And the Fool says: 'Yes indeed, thou woulds't make a good Foole'. And so did Hegel. On 1 January 1801, punctually, before the ink was dry on Hegel's dissertation, an eighth planet was discovered – the minor planet Ceres.

History has many ironies. The time-bomb in Gauss's curve is that after his death we discover that there is no God's eye view. The errors are inextricably bound up with the nature of human knowledge. And the irony is that the discovery was made in Göttingen.

Ancient university towns are wonderfully alike. Göttingen is like Cambridge in England or Yale in America: very provincial, not on the way to anywhere – no one comes to these backwaters except for the company of professors. And the professors are sure that this is the centre of the world. There is an inscription in the Rathskeller here which reads 'Extra Gottingam non est vita', 'Outside Göttingen there is no life'. This epigram, or should I call it epitaph, is not taken as seriously by the undergraduates as by the professors.

The symbol of the University is the iron statue outside the Rathskeller of a barefoot goosegirl that every student kisses at graduation. The University is a Mecca to which students come with something less than perfect faith. It is important that students bring a certain ragamuffin, barefoot irreverence to their studies; they are not here to worship what is known but to question it.

Like every university town, the Göttingen landscape is criss-crossed with long walks that professors take after lunch, and the research students are ecstatic if they are asked along. Perhaps Göttingen in the past had been rather sleepy. The small German university towns go back to a time before the country was united (Göttingen was founded by George II as ruler of Hanover), and this gives them a flavour of local bureaucracy. Even after the military might ended and the Kaiser abdicated in 1918, they were more conformist than universities outside Germany.

The link between Göttingen and the outside world was the railway. That was the way the visitors came from Berlin and abroad, eager to exchange the new ideas that were racing ahead in physics. It was a by-word in Göttingen that science came to life in the train to Berlin, because that is where people argued and contradicted and had new ideas. And had them challenged, too.

In the years of the First World War, science was dominated at Göttingen as elsewhere by Relativity. But in 1921 there was appointed to the chair of physics Max Born, who began a series of seminars that brought everyone interested in atomic physics here. It is rather surprising to reflect that Max Born was almost forty when he was appointed. By and large, physicists have done their best work before they are thirty (mathematicians even earlier, biologists perhaps a little later). But Born had a remarkable personal, Socratic gift. He drew young men to him, he got the best out of them, and the ideas that he and they exchanged and challenged also produced his best work. Out of that wealth of names, whom am I to choose? Obviously Werner Heisenberg, who did his finest work here with Born. Then, when Erwin Schrödinger published a different form of basic atomic physics, here is where the arguments took place. And from all over the world people came to Göttingen to join in.

It is rather strange to talk in these terms about a subject which, after all, is done by midnight oil. Did physics in the 1920s really consist of argument, seminar, discussion, dispute? Yes, it did. Yes, it still does. The people who met here, the people who meet in laboratories still, only end their work with a mathematical formulation. They begin it by trying to solve conceptual riddles. The riddles of the sub-atomic particles – of the electrons and the rest – are mental riddles.

Think of the puzzles that the electron was setting just at that time. The quip among professors was (because of the way university timetables are laid out) that on Mondays, Wednesdays, and Fridays the electron would behave like a particle; on Tuesdays, Thursdays, and Saturdays it would behave like a wave. How could you match those two aspects, brought from the large-scale world and pushed into a single entity, into this Lilliput, *Gulliver's Travels* world of the inside of the atom? That is what the speculation and argument was about. And that requires, not calculation, but insight, imagination – if you like, metaphysics. I remember a phrase that Max Born used when he came to England many years after, and that still stands in his autobiography. He said: 'I am now convinced that theoretical physics is actual philosophy'.

Max Born meant that the new ideas in physics amount to a different view of reality. The world is not a fixed, solid array of objects, out there, for it cannot be fully separated from our perception of it. It shifts under our gaze, it interacts with us, and the knowledge that it yields has to be interpreted by us. There is no way of exchanging information that does not demand an act of judgment. Is the electron a particle? It behaves like one in the Bohr atom. But de Broglie in 1924 made a beautiful wave model, in which the orbits are the places where an exact, whole number of waves closes round the nucleus. Max Born thought of a train of electrons as if each were riding on a crankshaft, so that collectively they constitute a series of Gaussian curves, a wave of probability. A new conception was being made, on the train to Berlin and the professorial walks in the woods of Göttingen: that whatever fundamental units the world is put together from, they are more delicate, more fugitive, more startling than we catch in the butterfly net of our senses.

All those woodland walks and conversations came to a brilliant climax in 1927. Early that year Werner Heisenberg gave a new characterisation of the electron. Yes, it is a particle, he said, but a particle which yields only limited information. That is, you can specify where it is at this instant, but then you cannot impose on it a specific speed and direction at the setting-off. Or conversely, if you insist that you are going to fire it at a certain speed in a certain direction, then you cannot specify exactly what its starting-point is – or, of course, its end-point.

That sounds like a very crude characterisation. It is not. Heisenberg gave it depth by making it precise. The information that the electron carries is limited in its totality. That is, for instance, its speed *and* its position fit *together* in such a way that they are confined by the tolerance of the quantum. This is the profound idea: one of the great scientific ideas, not only of the twentieth century, but in the history of science.

Heisenberg called this the Principle of Uncertainty. In one sense, it is a robust principle of the everyday. We know that we cannot ask the world to be exact. If an object (a familiar face, for example) had to be *exactly* the same before we recognised it, we would never recognise it from one day to the next. We recognise the object to be the same because it is much the same; it is never exactly like it was, it is tolerably like. In the act of recognition, a judgment is built in – an area of tolerance or uncertainty. So Heisenberg's principle says that no events, not even atomic events, can be described with certainty, that is, with zero tolerance. What makes the principle profound is that Heisenberg specifies the tolerance that can be reached. The measuring rod is Max Planck's quantum. In the world of the atom, the area of uncertainty is always mapped out by the quantum.

Yet the Principle of Uncertainty is a bad name. In science or outside it, we are not uncertain; our knowledge is merely confined within a certain tolerance. We should call it the Principle of Tolerance. And I propose that name in two senses. First, in the engineering sense. Science has progressed step by step, the most successful enterprise in the ascent of man, because it has understood that the exchange of information between man and nature, and man and man, can only take place with a certain tolerance. But second, I also use the word passionately about the real world. All knowledge, all information between human beings can only be exchanged within a play of tolerance. And that is true whether the exchange is in science, or in literature, or in religion, or in politics, or even in any form of thought that aspires to dogma. It is a major tragedy of my lifetime and yours that, here in Göttingen, scientists were refining to the most exquisite precision the Principle of Tolerance, and turning their backs on the fact that all around them tolerance was crashing to the ground beyond repair.

The sky was darkening all over Europe. But there was one particular cloud which had been hanging over Göttingen for a hundred years. Early in the 1800s Johann Friedrich Blumenbach had put together a collection of skulls that he got from distinguished gentlemen with whom he corresponded throughout Europe. There was no suggestion in Blumenbach's work that the skulls were to support a racist division of humanity, although he did use anatomical measurements to classify the families of man. All the same, from the time of Blumenbach's death in 1840 the collection was added to and added to and became a core of racist, pan-Germanic theory, which was officially sanctioned by the National Socialist Party when it came into power.

When Hitler arrived in 1933, the tradition of scholarship in Germany was destroyed, almost overnight. Now the train to Berlin was a symbol of flight. Europe was no longer hospitable to the imagination – and not just the scientific imagination. A whole conception of culture was in retreat: the conception that human knowledge is personal and responsible, an unending adventure at the edge of uncertainty. Silence fell, as after the trial of Galileo. The great men went out into a threatened world. Max Born. Erwin Schrödinger. Albert Einstein. Sigmund Freud. Thomas Mann. Bertolt Brecht. Arturo Toscanini. Bruno Walter. Marc Chagall. Enrico Fermi. Leo Szilard, arriving finally after many years at the Salk Institute in California.

The Principle of Uncertainty or, in my phrase, the Principle of Tolerance fixed once for all the realisation that all knowledge is limited. It is an irony of history that at the very time when this was being worked out there should rise, under Hitler in Germany and other tyrants elsewhere, a counter-conception: a principle of monstrous certainty. When the future looks back on the 1930s it will think of them as a crucial confrontation of culture as I have been expounding it, the ascent of man, against the throwback to the despots' belief that they have absolute certainty.

I must put all these abstractions into concrete terms, and I want to do so in one personality. Leo Szilard was greatly engaged in them,

Europe was no longer hospitable to the imagination.

Enrico Fermi.

and I spent many afternoons in the last year or so of his life talking with him about them at the Salk Institute.

Leo Szilard was a Hungarian whose university life was spent in Germany. In 1929 he had published an important and pioneer paper on what would now be called Information Theory, the relation between knowledge, nature and man. But by then Szilard was certain that Hitler would come to power, and that war was inevitable. He kept two bags packed in his room, and by 1933 he had locked them and taken them to England.

It happened that in September of 1933 Lord Rutherford, at the British Association meeting, made some remark about atomic energy never becoming real. Leo Szilard was the kind of scientist, perhaps just the kind of good-humoured, cranky man, who disliked any statement that contained the word 'never', particularly when made by a distinguished colleague. So he set his mind to think about the problem. He tells the story as all of us who knew him would picture it. He was living at the Strand Palace Hotel – he loved living in hotels. He was walking to work at Bart's Hospital, and as he came to Southampton Row he was stopped by a red light. (That is the only part of the story I find improbable; I never knew Szilard to stop for a red light.) However, before the light turned green, he had realised that if you hit an atom with one neutron, and it happens to break up and release two, then you would have a chain reaction. He wrote a specification for a patent which contains the words 'chain reaction' which was filed in 1934.

And now we come to a part of Szilard's personality which was characteristic of scientists at that time, but which he expressed most clearly and loudly. He wanted to keep the patent secret. He wanted to prevent science from being misused. And, in fact, he assigned the patent to the British Admiralty, so that it was not published until after the war.

But meanwhile war was becoming more and more threatening. The march of progress in nuclear physics and the march of Hitler went step by step, pace by pace, in a way that we forget now. Early in 1939 Szilard wrote to Joliot Curie asking him if one could make a prohibition on publication. He tried to get Fermi not to publish. But finally, in August of 1939, he wrote a letter which Einstein signed and sent to President Roosevelt, saying (roughly), 'Nuclear energy is

Albert Einstein
Old Grove Rd.
Nassau Point
Peconic, Long Island

August 2nd, 1939

F.D. Roosevelt,
President of the United States,
White House
Washington, D.C.

Sir:

Some recent work by E.Fermi and L. Szilard, which has been communicated to me in manuscript, leads me to expect that the element uranium may be turned into a new and important source of energy in the immediate future. Certain aspects of the situation which has arisen seem to call for watchfulness and, if necessary, quick action on the part of the Administration. I believe therefore that it is my duty to bring to your attention the following facts and recommendations:

In the course of the last four months it has been made probable - through the work of Joliot in France as well as Fermi and Szilard in America - that it may become possible to set up a nuclear chain reaction in a large mass of uranium,by which vast amounts of power and large quantities of new radium-like elements would be generated. Now it appears almost certain that this could be achieved in the immediate future.

This new phenomenon would also lead to the construction of bombs, and it is conceivable - though much less certain - that extremely powerful bombs of a new type may thus be constructed. A single bomb of this type, carried by boat and exploded in a port, might very well destroy the whole port together with some of the surrounding territory. However, such bombs might very well prove to be too heavy for transportation by air.

(*Above and Opposite*) Finally, Szilard wrote a letter which Einstein signed and sent to President Roosevelt.

Text of the letter of 2 August 1939 to the President of the United States.

-2-

The United States has only very poor ores of uranium in moderate quantities. There is some good ore in Canada and the former Czechoslovakia, while the most important source of uranium is Belgian Congo.

In view of this situation you may think it desirable to have some permanent contact maintained between the Administration and the group of physicists working on chain reactions in America. One possible way of achieving this might be for you to entrust with this task a person who has your confidence and who could perhaps serve in an inofficial capacity. His task might comprise the following:

a) to approach Government Departments, keep them informed of the further development, and put forward recommendations for Government action, giving particular attention to the problem of securing a supply of uranium ore for the United States;

b) to speed up the experimental work,which is at present being carried on within the limits of the budgets of University laboratories, by providing funds, if such funds be required, through his contacts with private persons who are willing to make contributions for this cause, and perhaps also by obtaining the co-operation of industrial laboratories which have the necessary equipment.

I understand that Germany has actually stopped the sale of uranium from the Czechoslovakian mines which she has taken over. That she should have taken such early action might perhaps be understood on the ground that the son of the German Under-Secretary of State, von Weizsäcker, is attached to the Kaiser-Wilhelm-Institut in Berlin where some of the American work on uranium is now being repeated.

Yours very truly,

A. Einstein

(Albert Einstein)

here. War is inevitable. It is for the President to decide what scientists should do about it'.

But Szilard did not stop. When in 1945 the European war had been won, and he realised that the bomb was now about to be made and used on the Japanese, Szilard marshalled protest everywhere he could. He wrote memorandum after memorandum. One memorandum to President Roosevelt only failed because Roosevelt died during the very days that Szilard was transmitting it to him. Always Szilard wanted the bomb to be tested openly before the Japanese and an international audience, so that the Japanese should know its power and should surrender before people died.

As you know, Szilard failed, and with him the community of scientists failed. He did what a man of integrity could do. He gave up physics and turned to biology – that is how he came to the Salk Institute – and persuaded others too. Physics had been the passion of the last fifty years, and their masterpiece. But now we knew that it was high time to bring to the understanding of life, particularly human life, the same singleness of mind that we had given to understanding the physical world.

The first atomic bomb was dropped on Hiroshima in Japan on 6 August 1945 at 8.15 in the morning. I had not been long back from Hiroshima when I heard someone say, in Szilard's presence, that it was the tragedy of scientists that their discoveries were used for destruction. Szilard replied, as he more than anyone else had the right to reply, that it was not the tragedy of scientists: 'it is the tragedy of mankind'.

There are two parts to the human dilemma. One is the belief that the end justifies the means. That push-button philosophy, that deliberate deafness to suffering, has become the monster in the war machine. The other is the betrayal of the human spirit: the assertion of dogma that closes the mind, and turns a nation, a civilisation, into a regiment of ghosts – obedient ghosts, or tortured ghosts.

It is said that science will dehumanise people and turn them into numbers. That is false, tragically false. Look for yourself. This is the concentration camp and crematorium at Auschwitz. This is where people were turned into numbers. Into this pond were flushed the ashes of some four million people. And that was not done by gas. It was done by arrogance. It was done by dogma. It was done by ignorance. When people believe that they have absolute knowledge, with no test in reality, this is how they behave. This is what men do when they aspire to the knowledge of gods.

Science is a very human form of knowledge. We are always at the brink of the known, we always feel forward for what is to be hoped. Every judgment in science stands on the edge of error, and is personal. Science is a tribute to what we can know although we are fallible. In the end the words were said by Oliver Cromwell: 'I beseech you, in the bowels of Christ, think it possible you may be mistaken'.

I owe it as a scientist to my friend Leo Szilard, I owe it as a human being to the many members of my family who died at Auschwitz, to stand here by the pond as a survivor and a witness. We have to cure ourselves of the itch for absolute knowledge and power. We have to close the distance between the push-button order and the human act. We have to touch people.

CHAPTER TWELVE

GENERATION UPON GENERATION

In the nineteenth century the city of Vienna was the capital of an Empire which held together a multitude of nations and languages. It was a famous centre of music, literature and the arts. Science was suspect in conservative Vienna, particularly biological science. But unexpectedly Austria was also the seedbed for one scientific idea (and in biology) that was revolutionary.

At the old university of Vienna the founder of genetics, and therefore of all the modern life sciences, Gregor Mendel, got such little university education as he had. He came at a historic time in the struggle between tyranny and freedom of thought. In 1848, shortly before he came, two young men had published far away in London, in German, a manifesto which begins with the phrase: 'Ein Gespenst geht um Europa', 'a spectre is haunting Europe', the spectre of communism.

Of course, Karl Marx and Friedrich Engels in *The Communist Manifesto* did not create the revolutions in Europe; but they gave them the voice. It was the voice of insurrection. A spate of disaffection ran though Europe against the Bourbons, the Habsburgs, and governments everywhere. Paris was in turmoil in February of 1848, and Vienna and Berlin followed. And so in the University Square in Vienna in March 1848 students protested and fought the police. The Austrian Empire, like others, shook. Metternich resigned and fled to London. The Emperor abdicated.

Emperors go, but empires remain. The new Emperor of Austria was a young man of eighteen, Franz Josef, who reigned like a medieval autocrat until the ramshackle empire fell to pieces during the First World War. I still remember Franz Josef when I was a small boy; like other Habsburgs, he had the long lower lip and pouched mouth which Velazquez has painted in the Spanish kings, and which is now recognised as a dominant genetic trait.

When Franz Josef came to the throne the patriots' speeches fell silent; the reaction under the young Emperor was total. At that moment the ascent of man was quietly set off in a new direction by the arrival at the University of Vienna of Gregor Mendel. He had been born Johann Mendel, a farmer's son; Gregor was the name he was given when he became a monk just before this, frustrated by poverty and lack of education. He remained all his life a farm boy in the way he went about his work, not a professor nor a gentleman naturalist like his contemporaries in England; he was a kitchen-garden naturalist.

Mendel had become a monk to get an education, and his abbot put him into the University of Vienna to get a formal diploma as a teacher. But he was nervous and was not a clever student. His examiner wrote that he 'lacks insight and the requisite clarity of knowledge' and failed him. The farm boy become monk had no choice except to withdraw again into the anonymity of the monastery at Brno in Moravia, which is now part of Czechoslovakia.

When Mendel came back from Vienna in 1853 he was, at the age of thirty-one, a failure. He had been sent by the Augustinian

The ascent of man was quietly set off in a new direction by Gregor Mendel.
Mendel in 1865.

Order of St Thomas in Brno, and they were a teaching order. The Austrian Government wanted the bright boys among the peasantry taught by monks. Theirs is the library not so much of a monastery as of a teaching order. And Mendel had failed to qualify as a teacher. He had to make up his mind whether to live the rest of his life as a failed teacher, or as – what? As the boy they called Hansl on the farm, the young man Johann from the farm, he decided; not as the monk Gregor. He went back in thought to what he had learned on the farm and had been fascinated by ever since: plants.

At Vienna he had been under the influence of the one fine biologist he ever met, Franz Unger, who took a concrete, practical view of inheritance: no spiritual essences, no vital forces, stick to the real facts. And Mendel decided to devote his life to practical experiments in biology, here in the monastery. A bold, silent, and secret stroke, I think, because the local bishop would not even allow the monks to teach biology.

Mendel began his formal experiments about two or three years after he came back from Vienna, say about 1856. He says in his paper that he worked for eight years. The plant that he had chosen, very carefully, is the garden pea. He picked out seven characters for comparison: shape of seed, colour of seed, and so on, finishing his list with tall in stem versus short-stemmed. And that last character is the one that I have chosen to display: tall versus short.

We do the experiment exactly throughout as Mendel did. We start by making a hybrid of tall and short, choosing the parent plants as Mendel specified:

> In experiments with this character, in order to be able to discriminate with certainty, the long axis of six to seven feet was always crossed with the short one of ¾ of a foot to 1½ feet.

In order to make sure that the short plant does not fertilise itself, we emasculate it. And then we artificially inseminate it from the tall plant.

The process of fertilisation takes its course. The pollen tubes grow down the ovules. The pollen nuclei (the equivalent of sperm in an animal) go down the pollen tubes and reach the ovules just as they do in any other fertilised pea. The plant bears pods that do not yet, of course, reveal their character.

The peas from the pods are now planted. Their development is at first indistinguishable from that of any other garden peas. But though they are only the first generation of hybrid offspring, their appearance when fully grown will already be a test of the traditional view of inheritance held by botanists then and long afterwards. The traditional view was that the characters of hybrids fall between the characters of their parents. Mendel's view was radically different, and he had even guessed a theory to explain it.

Mendel had guessed that a simple character is regulated by two particles (we now call them genes). Each parent contributes one of the two particles. If the two particles or genes are different, one will be dominant and the other recessive. The crossing of tall peas with short is a first step in seeing if this is true. And lo and behold, the first generation of hybrids, when fully grown, are all tall. In the language of modern genetics, the character tall is dominant over the character short. It is not true that the hybrids average the height of their parents; they are all tall plants.

Now the second step: we form the second generation as Mendel did. We fertilise the hybrids, this time with their own pollen. We allow the pods to form, plant the seeds, and here is the second generation. It is not all of anything, for it is not uniform; there is a majority of tall plants, but a significant minority of short plants. The fraction of the total that consists of short plants should be calculable from Mendel's guess about heredity; for if he was right, each hybrid in the first generation carried one dominant and one recessive gene. Therefore in one mating out of every four between first generation hybrids, two recessive genes have come together, and as a result one plant out of every four should be short. And so it is: in the second generation, one plant out of four is short, and three are tall. This is the famous ratio of one out of four, or one to three, that everyone associates with Mendel's name – and rightly so. As Mendel reported,

> Out of 1064 plants, in 787 cases the stem was long, and in 277 short. Hence a mutual ratio of 2.84 to 1 ... If now the results of the whole of the experiments be brought together, there is found, as between the number of forms with the dominant and recessive characters, an average ratio of 2.98 to 1, or 3 to 1.
>
> It is now clear that the hybrids form seeds having one or other of two differentiating characters, and of these one half develop again the hybrid form, while the other half yield plants which remain constant and receive the dominant or the recessive characters [respectively] in equal numbers.

Mendel published his results in 1866 in the *Journal of the Brno Natural History Society*, and achieved instant oblivion. No one cared. No one understood his work. Even when he wrote to a distinguished, rather stuffy figure in the field, Karl Nägeli, it was clear that he had no notion what Mendel was talking about. Of course, if Mendel had been a professional scientist, he would now have pushed to get the results known, and at least published the paper more widely in France or Britain in a journal that botanists and biologists read. He did try to reach scientists abroad by sending them reprints of his paper, but that is a long shot for an unknown writing in an unknown journal. However, at this moment, in 1868, two years after the paper was published, a most unexpected thing happened to Mendel. He was elected abbot of his monastery. And for the rest of his life he carried out his duties with commendable zeal, and a touch of neurotic punctilio.

He told Nägeli that he hoped to go on doing breeding experiments. But the only thing that Mendel now was able to breed were bees – he had always been anxious to push his work from plants to animals. And of course, being Mendel, he had his usual mixture of splendid intellectual fortune and practical bad luck. He made a hybrid strain of bees which gave excellent honey; but alas, they were so ferocious that they stung everybody for miles around and had to be destroyed.

Mendel seems to have been more exercised about tax demands on the monastery than about its religious leadership. And there is a hint that he was regarded as unreliable by the Emperor's Secret Police. Under the abbot's brow there lay a weight of private thought.

The puzzle of Mendel's personality is an intellectual one. No one could have conceived those experiments unless they had clearly in their minds the answer that they were going to get. It is a strange state of affairs, and I should give you chapter and verse for that.

First, a practical point. Mendel chose seven differences between peas to test for at the time, such as tall versus short, and so on. And indeed the pea does have seven pairs of chromosomes, so you can test for seven different characters in genes lying on seven different chromosomes. But that is the largest number you could have chosen. You could not test for eight different characters without getting two of the genes lying on the same chromosome, and therefore being at least partially linked. Nobody had thought of genes or heard of linkage then. Nobody had even heard of chromosomes at the time when Mendel was actually working on the paper.

Now surely you can be destined to be the abbot of a monastery, you can be chosen by God, but you cannot have that luck. Mendel must have done a good deal of observation and experiment before the formal work, in order to tease out these and convince himself that seven qualities or characters was just what he could get away with. There we glimpse the great iceberg of the mind in that secret, hidden face of Mendel's on which the paper and the achievement float. And you see it; you see it on every page of the manuscript – the algebraic symbolism, the statistics, the clarity of the exposition; everything is modern genetics, essentially as it is done now, but done more than a hundred years ago by an unknown.

And done by an unknown who had one crucial inspiration: that characters separate in an all-or-none fashion. Mendel conceived that in an age when biologists took it as axiomatic that crossing produces something between the two characters of the parents. We can hardly suppose that a recessive character never appeared, and we can only speculate that every time breeders observed this in a hybrid, it was thrown away because they were convinced that heredity must go by averaging.

Where did Mendel get the model of an all-or-nothing heredity? I think I know, but of course I cannot look into his head either. But there does exist one model (and it has existed since time immemorial) which is so obvious that perhaps no scientist would think of it: but a child, or a monk, might. That model is sex. Animals have been copulating for millions of years, and males and females of the same species do not produce sexual monsters or hermaphrodites: they produce either a male or a female. Men and women have been going to bed for upwards of a million years, at least; and they produce – what? Either men, or women. Some such simple, powerful model of an all-or-nothing way of passing on differences must have been in Mendel's mind, so that the experiments and the thought were clearly made for him of whole cloth, and fitted from the inception.

The monks, I think, knew this. I think they did not like what Mendel was doing. I think the bishop, who demurred at the peabreeding experiments, did not like it. They did not like his interest in the new biology at all – in Darwin's work, for example, which Mendel read and was impressed by. Of course, the routabout revolutionary Czech colleagues whom he often sheltered in the monastery were fond of him to the end. When he died in 1884, barely at the age of sixty-two, the great Czech composer Leoš Janáček played the organ at his funeral. But the monks elected a new abbot, and he burned all Mendel's papers at the monastery.

Mendel's great experiment remained forgotten for over thirty years until it was resurrected (by several scientists independently) in 1900. So his discoveries belong in effect to this century, when the study of genetics all at once blossoms from them.

To begin at the beginning. Life on earth has been going on for three thousand million years or more. For two-thirds of that time organisms reproduced themselves by cell division. Division produces identical offspring as a rule, and new forms appear only rarely, by mutation. For all that time, therefore, evolution was very slow. The first organisms to reproduce sexually were, it now seems, related to the green algae. That was less than a thousand million years ago. Sexual reproduction begins there, first in plants, then in animals. Since then its success has made it the biological norm, so that, for instance, we define two species as different if their members cannot breed with one another.

Sex produces diversity, and diversity is the propeller of evolution. The acceleration in evolution is responsible for the existence now of the dazzling variety of shape, colour, and behaviour in species. And we must count it also responsible for the proliferation of individual differences within species. All that was made possible by the emergence of two sexes. Indeed, the spread of sex through the biological world is itself a proof that species become fitted to a new environment by selection. For sex would not be necessary if the members of a species could inherit the acquired changes by which individuals adapt themselves. Lamarck at the end of the eighteenth century proposed that naive and, as it were, solitary mode of inheritance; but if it existed, it could be passed on better by cell division.

Two is the magic number. That is why sexual selection and courtship are so highly evolved in different species, in forms as spectacular as the peacock. It is why sexual behaviour is geared so precisely to the animal's environment. If the grunion could have adapted themselves without natural selection, then they would not trouble to dance on the Californian beaches in order to match incubation to the period of the moon. For them and for all the mavericks of adaptation, sex would not be necessary. And sex is itself a mode of natural selection of the fittest. Stags do not fight to kill, only to establish their right to choose the female. The multiplicity of shape, colour, and behaviour in individuals and in species is produced by the coupling of genes, as Mendel guessed. As a matter of mechanics, the genes are strung out along the chromosomes, which become visible only when the cell is dividing. But the question is not how the genes are arranged; the modern question is, How do they act? The genes are made of nucleic acids. *That* is where the action is.

How the message of inheritance is passed from one generation to the next was discovered in 1953, and it is the adventure story of science in the twentieth century. I suppose the moment of drama is the autumn of 1951, when a young man in his twenties, James Watson, arrives in Cambridge and teams up with a man of thirty-five, Francis Crick, to decipher the structure of deoxyribonucleic acid, DNA for

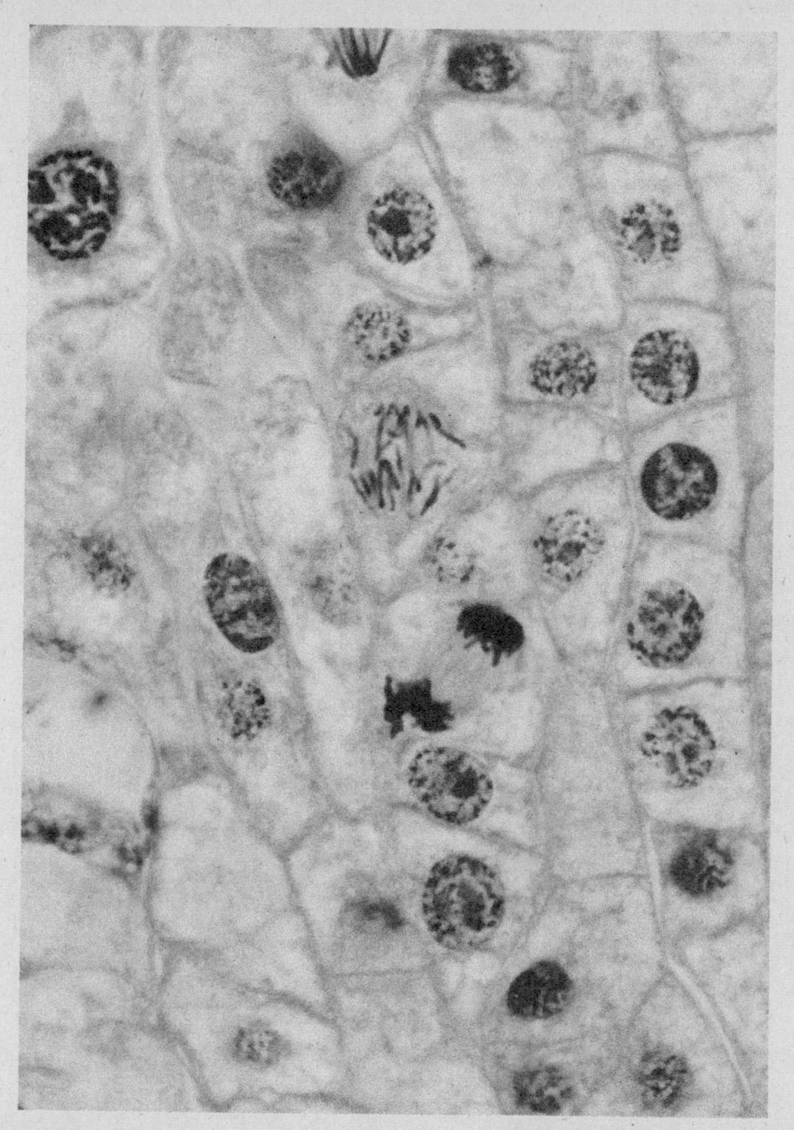

The genes are strung out along the chromosomes which become visible when the cell is dividing.

Large chromosomes of onion-skin cells.

short. DNA is a nucleic acid, that is, an acid in the central part of cells, and it had become clear in the preceding ten years that nucleic acids carry the chemical messages of inheritance from generation to generation. Two questions then faced the searchers in Cambridge, and in laboratories as far afield as California. What is the chemistry? And what is the architecture?

What is the chemistry? Which means, What are the parts that compose DNA and that can be shuffled about to make different forms of it? That was known pretty well. It was clear that DNA is made of sugars and phosphates (they were sure to be there, for structural reasons), and four specific small molecules or bases. Two of them are very small molecules, thymine and cytosine, in each of which atoms of carbon, nitrogen, oxygen, and hydrogen are arranged in a hexagon. And two of them are rather larger, guanine and adenine, in each of which the atoms are arranged in a hexagon and a pentagon joined together. It is usual in structural work to represent each of the small bases simply by a hexagon, and the large by the bigger figure, to draw attention to the shapes rather than the individual atoms.

And what is the architecture? Which means, What is the arrangement of the bases that gives DNA the ability to express many different genetic messages? For a building is not a heap of stones, and the DNA molecule is not a heap of bases. What gives it its structure and therefore its function? It was clear by then that the DNA molecule is a long extended chain, but rather rigid – a kind of organic crystal. And it seemed likely that it would be a helix (or spiral). How many helixes in parallel? One, two, three, four? There was a division of opinion into two main camps: the two-helix camp and the three-helix camp. And then, at the end of 1952, the great genius of structural chemistry, Linus Pauling, in California proposed a three-helix model. The backbone of sugar and phosphate ran down the middle, and the bases stuck out in all directions. Pauling's paper arrived in Cambridge in February 1953, and it was apparent to Crick and Watson that there was something wrong with it from the outset.

It may have been mere relief, it may have been a touch of malicious perversity, which made Jim Watson decide there and then that he would go for the double helix. After a visit to London,

> by the time I had cycled back to college and climbed over the back gate, I had decided to build two-chain models. Francis would have to agree. Even though he was a physicist, he knew that important biological objects come in pairs.

Moreover, he and Crick began to look for a structure with the backbones running on the outside: a sort of spiral staircase, with the sugars and phosphates holding it like two handrails. There were agonies of experimentation with cut-out shapes to see how the bases would fit as the steps in that model. And then, after one particularly wild mistake, all at once it became self-evident.

> I looked up, saw that it was not Francis, and began shifting the bases in and out of various other pairing possibilities. Suddenly I became aware that an adenine-thymine pair held together by two hydrogen bonds was identical in shape to a guanine-cytosine pair.

Of course: on each step there must be a small base and a large base. But not any large base. Thymine must be matched by adenine, and if you have cytosine then it must be matched by guanine. The bases go together in pairs of which each determines the other.

So the model of the DNA molecule is a spiral staircase. It is a right-handed spiral in which each tread is of the same size, at the same distance from the next, and turns at the same rate – thirty-six degrees between successive treads. And if cytosine is at one end of a tread, then guanine is at the other; and so for the other base pair. That implies that each half of the spiral carries the complete message, so that in a sense the other is redundant.

Let us build the molecule on a computer. Schematically, that is a base-pair; the dotted lines between the ends are the hydrogen bonds that hold the two bases together. We will put it into the end-on position in which we are going to stack it. And now we will stack it at the bottom of the left-hand side of the computer picture, where we are going to build the whole molecule of DNA, literally step by step.

Here is a second pair; it might be of the same kind as the first, or of the opposite kind; and it might face either way. We stack it over the first pair and turn it through thirty-six degrees. Here is a third pair, to which we do the same thing. And so on.

These treads are a code which will tell the cell step by step how to make the proteins necessary to life. The gene is forming visibly in front of our eyes, and the handrails of sugars and phosphates hold the spiral staircase rigid on each side. The spiral DNA molecule is a gene, a gene in action, and the treads are the steps by which it acts.

On 2 April 1953 James Watson and Francis Crick sent to *Nature* the paper which describes this structure in DNA on which they had worked for only eighteen months. In the words of Jacques Monod, of the Pasteur Institute in Paris and the Salk Institute in California,

> The fundamental biological invariant is DNA. That is why Mendel's defining of the gene as the unvarying bearer of hereditary traits, its chemical identification by Avery (confirmed by Hershey), and the elucidation by Watson and Crick of the structural basis of its replicative invariance, without any doubt

> constitute the most important discoveries ever made in biology. To which of course must be added the theory of natural selection, whose certainty and full significance were established only by those later discoveries.

The model of DNA patently lends itself to the process of replication which is fundamental to life even before sex. When a cell divides, the two spirals separate. Each base fixes opposite to it the other member of the pair to which it belongs. This is the point of the redundancy in the double helix: because each half carries the whole message or instruction, when a cell divides the same gene is reproduced. The magic number two here is the means by which a cell passes on its genetic identity when it divides.

The DNA spiral is not a monument. It is an instruction, a living mobile to tell the cell how to carry out the processes of life step by step. Life follows a time-table, and the treads of the DNA spiral encode and signal the sequence in which the time-table must go. The machinery of the cell reads off the treads in order, one after another. A sequence of three treads acts as a signal to the cell to make one amino acid. As the amino acids are formed in order, they line up and assemble in the cell as proteins. And the proteins are the agents and building blocks of life in the cell.

Every cell in the body carries the complete potential to make the whole animal, except only the sperm and egg cells. The sperm and the egg are incomplete, and essentially they are half cells: they carry half the total number of genes. Then when the egg is fertilised by the sperm, the genes from each come together in pairs as Mendel foresaw, and the total of instructions is assembled again. The fertilised egg is then a complete cell, and it is the model of every cell in the body. For every cell is formed by division of the fertilised egg, and is therefore identical with it in its genetic make-up. Like a chick embryo, the animal has the legacy of the fertilised egg all through life.

As the embryo develops the cells differentiate. Along the primitive streak the beginnings of the nervous system are laid down. Clumps of cells on either side will form the backbone. The cells specialise: nerve cells, muscle cells, connective tissue (the ligaments and tendons), blood cells, blood vessels. The cells specialise because they have accepted the DNA instructions to make the proteins that are appropriate to the functioning of that cell and no other. This is the DNA in action.

The baby is an individual from birth. The coupling of genes from both parents stirred the pool of diversity. The child inherits gifts from both parents, and chance has now combined these gifts in a new and original arrangement. The child is not a prisoner of its inheritance; it holds its inheritance as a new creation which its future actions will unfold.

The child is an individual. The bee is not, because the drone bee is one of a series of identical replicas. In any hive the queen is the only fertile female. When she mates with a drone in mid-air, she goes on hoarding his sperms; the drone dies. If the queen now releases a sperm with an egg she lays, she makes a worker bee, a female. If she lays an egg but releases no sperm with it, a drone bee is made, a male, in a sort of virgin birth. It is a totalitarian paradise, for ever loyal, for ever fixed, because it has shut itself off from the adventure of diversity that drives and changes the higher animals and man.

A world as rigid as the bee's could be created among higher animals, even among men, by cloning: that is, by growing a colony or clone of identical animals from cells of a single parent. Begin with a mixed population of an amphibian, the axolod. Suppose we decided to fix on one type, the speckled axolotl. We take some eggs from a speckled female and grow an embryo which is destined to be speckled. Now we tease out from the embryo a number of cells. Wherever in the embryo we take them from, they are identical in their genetic make-up, and each cell is capable of growing into a complete animal – our procedure will prove that.

We are going to grow identical animals, one from each cell. We need a carrier in which to grow the cells: any axolotl carrier will do – she can be white. We take unfertilised eggs from the carrier and destroy the nucleus in each egg. And into it we insert one of the single identical cells of the speckled parent of the clone. These eggs will now grow into speckled axolotls.

The clone of identical eggs made in this way are all grown at the same time. Each egg divides at the same moment – divides once, divides twice, and goes on dividing. All that is normal, exactly as in any egg. At the next stage, single cell divisions are no longer visible. Each egg has turned into a kind of tennis ball, and begins to turn itself inside out – or it would be more literal to say, outside in. Still all the eggs are in step. Each egg folds over to form the animal, always in step: a regimented world in which the units obey every command identically at the identical moment, except (we see) one unfortunate that has been deprived and is falling behind. And finally we have the clone of individual axolotls, each of them an identical copy of the parent, and each of them a virgin birth like the drone bee.

Should we make clones of human beings – copies of a beautiful mother, perhaps, or of a clever father? Of course not. My view is that diversity is the breath of life, and we must not abandon that for any single form which happens to catch our fancy – even our genetic fancy. Cloning is the stabilisation of one form, and that runs against the whole current of creation – of human creation above all. Evolution is founded in variety and creates diversity; and of all animals, man is most creative because he carries and expresses the largest store of variety. Every attempt to make us uniform, biologically, emotionally, or intellectually, is a betrayal of the evolutionary thrust that has made man its apex.

Yet it is odd that the myths of creation in human cultures seem almost to yearn back for an ancestral clone. There is a strange suppression of sex in the ancient stories of origins. Eve is cloned from Adam's rib, and there is a preference for virgin birth.

Happily we are not frozen into identical copies. In the human species sex is highly developed. The female is receptive at all times, she has permanent breasts, she takes an active part in sexual selection. Eve's apple, as it were, fertilises mankind; or at least spurs it to its ageless preoccupation.

It is obvious that sex has a very special character for human beings. It has a special biological character. Let us take one simple, down-to-earth criterion for that: we are the only species in which the female has orgasms. That is remarkable, but it is so. It is a mark of the fact that in general there is much less difference between men and women (in the biological sense and in sexual behaviour) than there is in other species. That may seem a strange thing to say. But to the gorilla and the chimpanzee, where there are enormous differences between male and female, it would be obvious. In the language of biology, sexual dimorphism is small in the human species.

So much for biology. But there is a point on the borderline between biology and culture which really marks the symmetry in sexual behaviour, I think, very strikingly. It is an obvious one. We are the only species that copulates face to face, and this is universal in all cultures. To my mind, it is an expression of a general equality which has been important in the evolution of man, I think, right back to the time of *Australopithecus* and the first tool-makers.

Why do I say that? Well, we have something to explain. We have to explain the speed of human evolution over a matter of one,

Eve is cloned from Adam's rib.
'The Creation of Eve' by Andrea Pisano.

three, let us say five million years at most. That is terribly fast. Natural selection simply does not act as fast as that on animal species. We, the hominids, must have supplied a form of selection of our own; and the obvious choice is sexual selection. There is evidence now that women marry men who are intellectually like them, and men marry women who are intellectually like them. And if that preference really goes back

(*Opposite*) Sex was invented as a biological instrument by the algae.

Cell of the green alga, Spirogyra, in the process of fusion. Ancestors of this species produced the first evidence of cells fusing to form fertile egg cells.

over some million of years, then it means that selection for skills has always been important on the part of both sexes.

I believe that as soon as the forerunners of man began to be nimble with their hands in making tools and clever with their brains in planning them, the nimble and clever enjoyed a selective advantage. They were able to get more mates and to beget and feed more children than the rest. If my speculation about this is right, it explains how the nimble-fingered and quick-witted were able to dominate the biological evolution of man, and take it ahead so fast. And it shows that even in his biological evolution, man has been nudged and driven by a cultural talent, the ability to make tools and communal plans. I think that is still expressed in the care that kindred and community take in all cultures, and only in human cultures, to arrange what is revealingly called a good match.

Yet if that had been the only selective factor then, of course, we should be much more homogeneous than we are. What keeps alive the variety among human beings? That is a cultural point. In every culture there are also special safeguards to make for variety. The most striking of them is the universal prohibition of incest (for the man in the street – it does not always apply to royal families). The prohibition of incest only has a meaning if it is designed to prevent older males dominating a group of females, as they do in (let us say) ape groups.

The preoccupation with the choice of a mate both by male and female I regard as a continuing echo of the major selective force by which we have evolved. All that tenderness, the postponement of marriage, the preparations and preliminaries that are found in all cultures, are an expression of the weight that we give to the hidden qualities in a mate. Universals that stretch across all cultures are rare and tell-tale. Ours is a cultural species, and I believe that our unique attention to sexual choice has helped to mould it.

Most of the world's literature, most of the world's art, is preoccupied with the theme of boy meets girl. We tend to think of this as being a sexual preoccupation that needs no explanation. But I think that is a mistake. On the contrary, it expresses the deeper fact that we are uncommonly careful in the choice, not of whom we take to bed, but by whom we are to beget children. Sex was invented as a biological instrument by (say) the green algae. But as an instrument in the ascent of man which is basic to his cultural evolution, it was invented by man himself

Spiritual and carnal love are inseparable. A poem by John Donne says that; he called it *The Extasie*, and I quote eight lines from almost eighty.

> All day, the same our postures were,
> And wee said nothing, all the day.
>
> But O alas, so long, so farre,
> Our bodies why doe wee forbeare?
>
> This Extasic doth unperplex
> (We said) and tell us what we love.
>
> Loves mysteries in soules doe grow,
> But yet the body is his booke.

CHAPTER THIRTEEN

THE LONG CHILDHOOD

I begin this last essay in Iceland because it is the seat of the oldest democracy in Northern Europe. In the natural amphitheatre of Thingvellir, where there were never any buildings, the Allthing of Iceland (the whole community of the Norsemen of Iceland) met each year to make laws and to receive them. And this began about AD 900, before Christianity arrived, at a time when China was a great empire, and Europe was the spoil of princelings and robber barons. That is a remarkable beginning to democracy.

But there is something more remarkable about this misty, inclement site. It was chosen because the farmer who had owned it had killed, not another farmer but a slave, and had been outlawed. Justice was seldom so even-handed in slave-owning cultures. Yet justice is a universal of all cultures. It is a tightrope that man walks, between his desire to fulfil his wishes, and his acknowledgement of social responsibility. No animal is faced with this dilemma: an animal is either social or solitary. Man alone aspires to be both in one, a social solitary. And to me that is a unique biological feature. That is the kind of problem that engages me in my work on human specificity, and that I want to discuss.

It is something of a shock to think that justice is part of the biological equipment of man. And yet it is exactly that thought which took me out of physics into biology, and that has taught me since that a man's life, a man's home, is a proper place in which to study his biological uniqueness.

It is natural that by tradition biology is thought of in a different way: that the likeness between man and the animals is what dominates it. Back before the year AD 200 the great classic author of antiquity in medicine, Claudius Galen, studied, for example, the forearm in man. How did he study it? By dissecting the forearm in a Barbary ape. That is how you have to begin, necessarily using the evidence of the animals, long before the theory of evolution comes to justify the analogy. And to this day the wonderful work on animal behaviour by Konrad Lorenz naturally makes us seek for likeness between the duck and the tiger and man; or B. F. Skinner's psychological work on pigeons and rats. They tell us something about man. But they cannot tell us everything. There must be something unique about man because otherwise, evidently, the ducks would be lecturing about Konrad Lorenz, and the rats would be writing papers about B. F. Skinner.

Let us not beat about the bush. The horse and the rider have many anatomical features in common. But it is the human creature that rides the horse, and not the other way about. And the rider is a very good example, because man was not created to ride the horse. There is no wiring inside the brain that makes us horse riders. Riding a horse is a comparatively recent invention, less than five thousand years old. And yet it has had an immense influence, for instance on our social structure.

The plasticity of human behaviour makes that possible. That is what characterises us; in our social institutions, of course, but for me, naturally, above all in books, because they are the permanent product of the total interests of the human mind. They come to me like the memory of my parents: Isaac Newton, the great man dominating the Royal Society at the beginning of the eighteenth century, and William

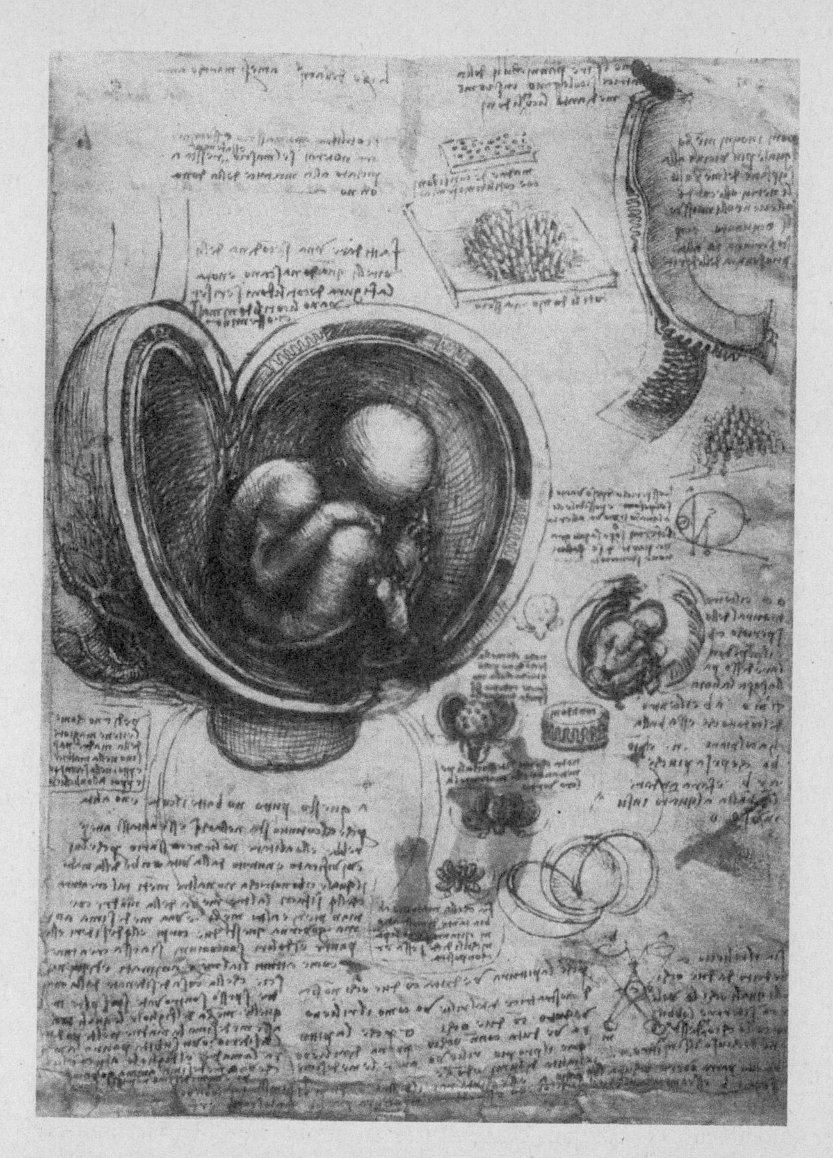

The brain and the baby is exactly where the plasticity of human behaviour begins.

Leonardo da Vinci's anatomical notes on the human foetus.

Blake, writing the *Songs of Innocence* late in the eighteenth century. They are two aspects of the one mind, and both are what behavioural biologists call species-specific.

How can I put this most simply? I wrote a book recently called *The Identity of Man*. I never saw the cover of the English edition until the book reached me in print. And yet the artist had understood exactly what was in my mind, by putting on the cover a drawing of the brain and the *Mona Lisa*, one on top of the other. In his action he demonstrated what the book said. Man is unique not because he does science, and he is unique not because he does art, but because science and art equally are expressions of his marvellous plasticity of mind. And the *Mona Lisa* is a very good example, because after all what did Leonardo do for much of his life? He drew anatomical pictures, such as the baby in the womb in the Royal Collection at Windsor. And the brain and the baby is exactly where the plasticity of human behaviour begins.

I have an object which I treasure: a cast of the skull of a child that is two million years old, the Taung baby. Of course, it is not strictly a human child. And yet if she – I always think of her as a girl – if she had lived long enough, she might have been my ancestor. What distinguishes her little brain from mine? In a simple sense, the size. That brain, if she had grown up, would have weighed perhaps a little over a pound. And my brain, the average brain today, weighs three pounds.

I am not going to talk about the neural structures, about one-way conduction in nervous tissues, or even about the old brain and the new, because that apparatus is what we share with many animals. I am going to talk about the brain as it is specific to the human creature.

The first question we ask is, is the human brain a better computer – a more complex computer? Of course, artists in particular tend to

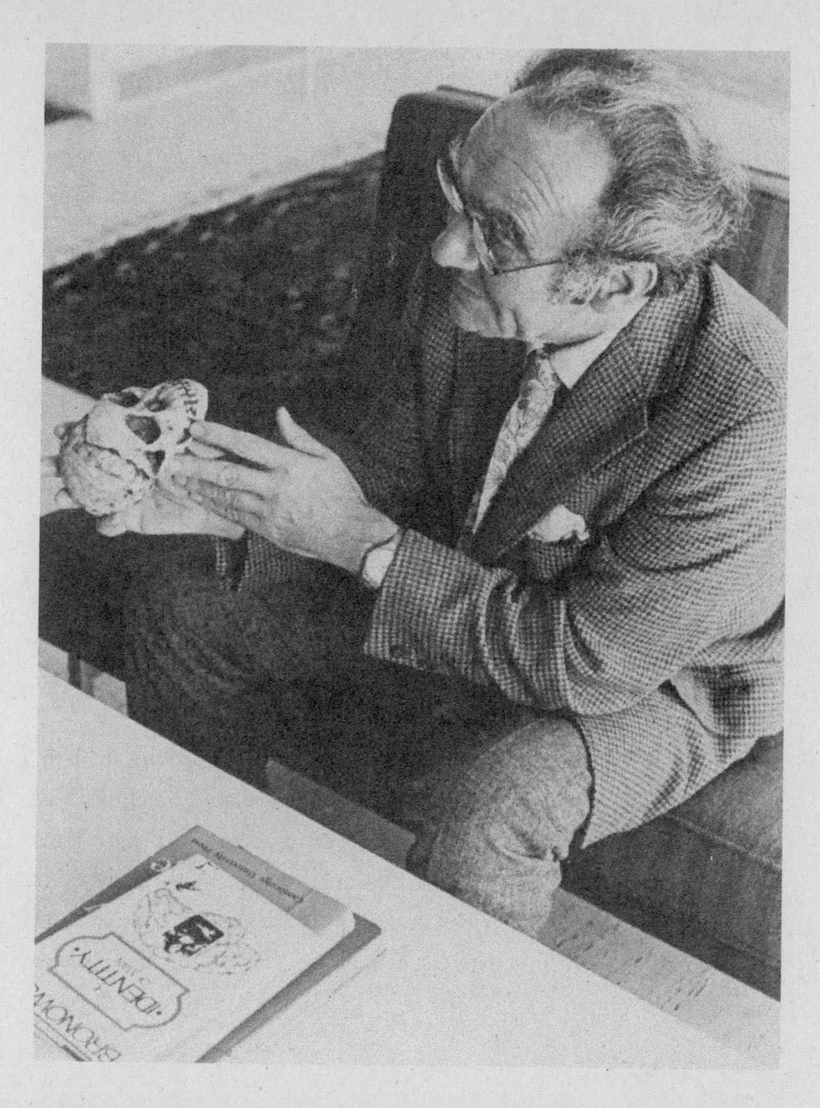

Man is unique not because he does science, and not because he does art, but because science and art equally are expressions of his marvellous plasticity of mind.

The author at home, with an endocast of the Taung child's skull. A copy of his book The Identity of Man *is on the table. La Jolla, California, 1973.*

think of the brain as a computer. So in his *Portrait of Dr Bronowski* Terry Durham has symbols of the spectrum and the computer, because that is how an artist imagines a scientist's brain. But of course that cannot be right. If the brain were a computer, then it would be carrying out a pre-wired set of actions in an inflexible sequence.

By way of example, think of a very beautiful piece of animal behaviour described in my friend Dan Lehrman's work on the mating of the ring-dove. If the male coos in the right way, if he bows in the right way, then the female explodes in excitement, all her hormones squirt, and she goes through a sequence as part of which she builds a perfect nest. Her actions are exact in detail and order, yet they are untaught, and therefore invariable; the ring-dove never changes them. Nobody ever gave her any set of bricks to learn to build a nest. But you could not get a human being to build anything unless the child had put together a set of bricks. That is the beginning of the Parthenon and the Taj Mahal, of the dome at Sultaniyeh and the Watts Towers, of Machu Picchu and the Pentagon.

We are not a computer that follows routines laid down at birth. If we are any kind of machine, then we are a learning machine, and we do our important learning in specific areas of the brain. Thus you see that the brain has not just blown up to two or three times its size during its evolution. It has grown in quite special areas: where it controls the hand, for instance, where speech is controlled, where foresight and planning are controlled. I shall ask you to look at them one by one.

Consider the hand first. The recent evolution of man certainly begins with the advancing development of the hand, and the selection for a brain which is particularly adept at manipulating the hand. We feel the pleasure of that in our actions, so that for the artist the hand remains a major symbol: the hand of Buddha, for instance,

Only man can oppose the thumb precisely to the forefinger.
Self-portrait, Albrecht Dürer.

giving man the gift of humanity in a gesture of calm, the gift of fearlessness. But also for the scientist the hand has a special gesture: we can oppose the thumb to the fingers. Well, the apes can do that. But we can oppose the thumb precisely to the forefinger, and that is a special human gesture. And it can be done because there is an area in the brain so large that I can best describe its size to you in the following way: we spend more grey matter in the brain manipulating the thumb than in the total control of the chest and the abdomen.

I remember as a young father tiptoeing to the cradle of my first daughter when she was four or five days old, and thinking, 'These marvellous fingers, every joint so perfect, down to the finger nails. I could not have designed that detail in a million years'. But of course it is exactly a million years that it took me, a million years that it took mankind, for the hand to drive the brain and for the brain to feed back and drive the hand to reach its present stage of evolution. And that takes place in a quite specific place in the brain. The whole of the hand is essentially monitored by a part of the brain that can be marked out, near the top of the head.

Take next an even more specifically human part of the brain which does not exist in animals at all: for speech. That is localised in two connected areas of the human brain; one area is close to the hearing centre, and the other lies forward and higher, in the frontal lobes. Is that pre-wired? Yes, in one sense, because if we do not have the speech centres intact we cannot speak at all. And yet, does it have to be learned? Of course it does. I speak English, which I only learned at the age of thirteen; but I could not speak English if I had not before learned language. You see, if you leave a child speaking no language until the age of thirteen, then it is almost impossible for it to learn at all. I speak English because I learned Polish at the age of two. I have forgotten every word of Polish, but I learned *language*. Here as in other human gifts the brain is wired to learn.

The speech areas are very peculiar in another way that is human. You know that the human brain is not symmetrical in its two halves.

The evidence is familiar to you in the observation that, unlike other animals, men are markedly right-handed or left-handed. Speech also is controlled on one side of the brain, but the side does not vary. Whether you are right-handed or left-handed, speech is almost certainly on the left. There are exceptions, in the same way that there are people who have their heart on the right, but the exceptions are rare: by and large speech is in areas in the left half of the brain. And what is in the matching areas on the right? We do not exactly know, so far. We do not exactly know what the right-hand side of the brain does in those areas which are devoted to speech on the left. But it looks as if they take the input that comes by way of the eye – the map of a two-dimensional world on the retina – and turn it or organise it into a three-dimensional picture. If that is right, then in my view it is clear that speech is also a way of organising the world into its parts and putting them together again like movable images.

The organisation of experience is very far-sighted in man, and is lodged in a third area of human specificity. The main organisation of the brain is in the frontal lobes and the prefrontal lobes. I am, every man is, a highbrow, an egghead, because that is how his brain goes. By contrast, we know that the Taung skull is not just that of a child that died recently and that we have mistaken for a fossil, because she still has a rather sloping forehead.

Exactly what do these large frontal lobes do? They may well have several functions, certainly, and yet do one very specific and important thing. They enable you to think of actions in the future, and wait for a reward then. Some beautiful experiments on this delayed response were first done by Walter Hunter round about 1910, and then refined by Jacobsen in the 1930s. The kind of thing that Hunter did was this: he would take some reward, and he would show it to an animal and then hide it. The results found in the darling of the laboratory, the rat, are typical. If you take a rat and, having shown it the reward, you let it go at once, the rat of course goes to the hidden reward immediately. But if you keep the rat waiting for some minutes, then it is no longer able to identify where it ought to go for its reward.

Of course, children are quite different. Hunter did the same experiments with children, and you can keep children of five or six waiting for half an hour, perhaps an hour. Hunter had a little girl whom he was trying to keep amused while keeping her waiting, and he talked to her. Finally she said to him, 'You know, I think you're just trying to make me forget'.

The ability to plan actions for which the reward is a long way off is an elaboration of the delayed response, and sociologists call it 'the postponement of gratification'. It is a central gift that the human brain has to which there is no rudimentary match in animal brains until they become quite sophisticated, well up in the evolutionary scale, like our cousins the monkeys and the apes. That human development means that we are concerned in our early education actually with the postponement of decisions. Here I am saying something different from the sociologists. We have to put off the decision-making process, in order to accumulate enough knowledge as a preparation for the future. That seems an extraordinary thing to say. But that is what childhood is about, that is what puberty is about, that is what youth is about.

I want to put my stress on the postponement of *decision* quite dramatically – and I mean the word literally. What is the major drama in the English language? It is *Hamlet*. What is *Hamlet* about? It is a play about a young man – a boy – who is faced with the first great decision of his life. And it is a decision beyond his reach: to kill the murderer of his father. It is pointless of the Ghost to keep on nudging him and saying, 'Revenge, Revenge'. The fact is that Hamlet as a youth is simply not mature. Intellectually or emotionally, he is not ripe for the act that he is asked to perform. And the whole play is an endless postponement of his decision while wrestling with himself.

The high point is in the middle of Act III. Hamlet sees the King at prayer. The stage directions are so uncertain that he may even hear the King at prayer, confessing his crime. And what does Hamlet say? 'Now might I do it – pat!' But he does not do it; he is simply not ready for an act of that magnitude in boyhood. So at the end of the play Hamlet is murdered. But the tragedy is not that Hamlet dies; it is that he dies exactly when he is ready to become a great king.

In man, before the brain is an instrument for action, it has to be an instrument of preparation. For that, quite specific areas are involved; for example, the frontal lobes have to be undamaged. But, far more deeply, it depends on the long preparation of human childhood.

In scientific terms we are neotenous; that is, we come from the womb still as embryos. And perhaps that is why our civilisation, our scientific civilisation, adores above all else the symbol of the child, ever since the Renaissance: the Christ child painted by Raphael and re-enacted by Blaise Pascal; the young Mozart and Gauss; the children in Jean Jacques Rousseau and Charles Dickens. It never struck me that other civilisations are different until I sailed south from here out of California, four thousand miles away to Easter Island. There I was struck by the historical difference.

Every so often some visionary invents a new Utopia: Plato, Sir Thomas More, H. G. Wells. And always the idea is that the heroic image shall last, as Hitler said, for a thousand years. But the heroic images always look like the crude, dead, ancestral faces of the statues on Easter Island – why, they even look like Mussolini! That is not the essence of the human personality, even in terms of biology. Biologically, a human being is changeable, sensitive, mutable, fitted to many environments, and not static. The real vision of the human being is the child wonder, the Virgin and Child, the Holy Family.

When I was a boy in my teens, I used to walk on Saturday afternoons from the East End of London to the British Museum, in order to look at the single statue from the Easter Islands which somehow they had not got inside the Museum. So I am fond of these ancient ancestral faces. But in the end, all of them are not worth one child's dimpled face.

If I was a little carried away in saying that at Easter Island, it was with reason. Think of the investment that evolution has made in the child's brain. My brain weighs three pounds, my body weighs fifty times as much as that. But when I was born, my body was a mere appendage to the head; it weighed only five or six times as much as my brain. For most of history, civilisations have crudely ignored that enormous potential. In fact the longest childhood has been that of civilisation, learning to understand that.

For most of history, children have been asked simply to conform to the image of the adult. We travelled with the Bakhtiari of Persia on their spring migration. They are as near as any surviving, vanishing people can be to the nomad ways of ten thousand years ago. You see it everywhere in such ancient modes of life: the image of the adult shines in the children's eyes. The girls are little mothers in the making. The boys are little herdsmen. They even carry themselves like their parents.

History, of course, did not stand still between the nomad and the Renaissance. The ascent of man has never come to a stop. But the ascent of the young, the ascent of the talented, the ascent of the imaginative: that became very halting many times in between.

Of course there were great civilisations. Who am I to belittle the civilisations of Egypt, of China, of India, even of Europe in the Middle

Ages? And yet by one test they all fail: they limit the freedom of the imagination of the young. They are static, and they are minority cultures. Static, because the son does what the father did, and the father what the grandfather did. And minority, because only a tiny fraction of all that talent that mankind produces is actually used; learns to read, learns to write, learns another language, and climbs the terribly slow ladder of promotion.

In the Middle Ages the ladder of promotion was through the Church; there was no other way for a clever, poor boy to go up. And at the end of the ladder there is always the image, the icon of the godhead that says, 'Now you have reached the last commandment: Thou shalt not question'.

For instance, when Erasmus was left an orphan in 1480, he had to prepare for a career in the Church. The services were as beautiful then as now. Erasmus may himself have taken part in the moving Mass *Cum Giubilate* of the fourteenth century, which I have heard in a church that is even older, San Pietro in Gropina. But the monk's life was for Erasmus an iron door closed against knowledge. Only when Erasmus read the classics for himself, in defiance of orders, did the world open for him. 'A heathen wrote this to a heathen,' he said, 'yet it has justice, sanctity, truth. I can hardly refrain from saying "Saint Socrates, pray for me!"'

Erasmus made two lifelong friends, Sir Thomas More in England and Johann Frobenius in Switzerland. From More he got what I got when I first came to England, the sense of pleasure in the companionship of civilised minds. From Frobenius he got a sense of the power of the printed book. Frobenius and his family were the great printers of the classics in the 1500s, including the classics of medicine. Their edition of the works of Hippocrates is, I think, one of the most beautiful books ever printed, in which the happy passion of the printer sits on the page as powerful as the knowledge.

What did those three men and their books mean – the works of Hippocrates, More's *Utopia, The Praise of Folly* by Erasmus? To me, this is the democracy of the intellect; and that is why Erasmus and Frobenius and Sir Thomas More stand in my mind as gigantic landmarks of their time. The democracy of the intellect comes from the printed book, and the problems that it set from the year 1500 have lasted right down to the student riots of today. What did Sir Thomas More die of? He died because his king thought of him as a wielder of power. And what More wanted to be, what Erasmus wanted to be, what every strong intellect wants to be, is a guardian of integrity.

There is an age-old conflict between intellectual leadership and civil authority. How old, how bitter, came home to me when I came up from Jericho on the road that Jesus took, and saw the first glimpse of Jerusalem on the skyline as he saw it going to his certain death. Death, because Jesus was then the intellectual and moral leader of his people, but he was facing an establishment in which religion was simply an arm of government. And that is a crisis of choice that leaders have faced over and over again: Socrates in Athens; Jonathan Swift in Ireland, torn between pity and ambition; Mahatma Gandhi in India; and Albert Einstein, when he refused the presidency of Israel.

I bring in the name of Einstein deliberately because he was a scientist, and the intellectual leadership of the twentieth century rests with scientists. And that poses a grave problem, because science is also a source of power that walks close to government and that the state wants to harness. But if science allows itself to go that way, the beliefs of the twentieth century will fall to pieces in cynicism. We shall be left without belief, because no beliefs can be built up in this century that are not based on science as the

There is an age-old conflict between intellectual leadership and civil authority. How old, how bitter, came home to me when I saw my first glimpse of Jerusalem on the skyline of the road up from Jericho.
A panoramic view of the old city of Jerusalem, Israel.

recognition of the uniqueness of man, and a pride in his gifts and works. It is not the business of science to inherit the earth, but to inherit the moral imagination; because without that man and beliefs and science will perish together.

I must bring that concretely into the present. The man who personifies these issues for me is John von Neumann. He was born in 1903, the son of a Jewish family in Hungary. If he had been born a hundred years earlier, we would never have heard of him. He would have been doing what his father and grandfather did, making rabbinical comments on dogma.

Instead, he was a child prodigy of mathematics, 'Johnny' to the end of his life. In his teens he already wrote mathematical papers. He did the great work on both the subjects for which he is famous before he was twenty-five.

Both subjects are concerned, I suppose I should say, with play. You must see that in a sense all science, all human thought, is a form of play. Abstract thought is the neoteny of the intellect, by which man is able to continue to carry out activities which have no immediate goal (other animals play only while young) in order to prepare himself for long-term strategies and plans.

I worked with Johnny von Neumann during the Second World War in England. He first talked to me about his *Theory of Games* in a taxi in London – one of the favourite places in which he liked to talk about mathematics. And I naturally said to him, since I am an enthusiastic chess player, 'You mean, the theory of games like chess.' 'No, no,' he said. 'Chess is not a game. Chess is a well-defined form of computation. You may not be able to work out the answers, but in theory there must be a solution, a right procedure in any position. Now real games', he said, 'are not like that at all. Real life is not like that. Real life consists of bluffing, of little tactics of deception, of asking yourself what is the other man going to think I mean to do. And that is what games are about in my theory.'

And that is what his book is about. It seems very strange to find a book, large and serious, entitled the *Theory of Games and Economic Behavior*, in which there is a chapter called 'Poker and Bluffing'. How surprising and how forbidding, moreover, to find it covered with equations that look so very pompous. Mathematics is not a pompous activity, least of all in the hands of extraordinarily fast and penetrating minds like Johnny von Neumann. What is running through the page is a clear intellectual line like a tune, and all the heavy weight of equations is simply the orchestration down in the bass.

In the latter part of his life, John von Neumann carried this subject into what I call his second great creative idea. He realised that computers would be technically important, but he also began to realise that one must understand clearly how real-life situations are different from computer situations, exactly because they do not have the precise solutions that chess or engineering calculations do.

I will use my own terms to describe John von Neumann's achievement, instead of his technical ones. He distinguished between short-term tactics and grand, long-term strategies. Tactics can be calculated exactly, but strategies cannot. Johnny's mathematical and conceptual success was in showing that nevertheless there are ways to form best strategies.

And in his last years he wrote a beautiful book called *The Computer and the Brain*, the Silliman Lectures that he should have given, but was too ill to give, in 1956. In them he looks at the brain as having a language in which the activities of the different parts of the brain have somehow to be interlocked and made to match so that we devise a plan, a procedure, as a grand overall way of life – what in the humanities we would call a system of values.

There was something endearing and personal about Johnny von Neumann. He was the cleverest man I ever knew, without exception. And he was a genius, in the sense that a genius is a man who has great ideas. When he died in 1957 it was a great tragedy to us all. And that was not because he was a modest man. When I worked with him during the war, we once faced a problem together, and he said to me at once, 'Oh no, no, you are not seeing it. Your kind of visualising mind is not right for seeing this. Think of it abstractly. What is happening on this photograph of an explosion is that the first differential coefficient vanishes identically, and that is why what becomes visible is the trace of the second differential coefficient.'

As he said, that is not the way I think. However, I let him go to London. I went off to my laboratory in the country. I worked late into the night. Round about midnight I had his answer. Well, John von Neumann always slept very late, so I was kind and I did not wake him until well after ten in the morning. When I called his hotel in London, he answered the phone in bed, and I said, 'Johnny, you're quite right.' And he said to me, 'You wake me up early in the morning to tell me that I'm right? Please wait until I'm wrong.'

If that sounds very vain, it was not. It was a real statement of how he lived his life. And yet it has something in it which reminds me that he wasted the last years of his life. He never finished the great work that has been very difficult to carry on since his death. And he did not, really because he gave up asking himself how other *people* see things. He became more and more engaged in work for private firms, for industry, for government. They were enterprises which brought him to the centre of power, but which did not advance either his knowledge or his intimacy with people – who to this day have not yet got the message of what he was trying to do about the human mathematics of life and mind.

Johnny von Neumann was in love with the aristocracy of intellect. And that is a belief which can only destroy the civilisation that we know. If we are anything, we must be a democracy of the intellect. We must not perish by the distance between people and government, between people and power, by which Babylon and Egypt and Rome failed. And that distance can only be conflated, can only be closed, if knowledge sits in the homes and heads of people with no ambition to control others, and not up in the isolated seats of power.

That seems a hard lesson. After all, this is a world run by specialists: is not that what we mean by a scientific society? No, it is not. A scientific society is one in which specialists can indeed do the things like making the electric light work. But it is you, it is I, who have to know how nature works, and how (for example) electricity is one of her expressions in the light and in my brain.

We have not advanced the human problems of life and mind that once occupied John von Neumann. Will it be possible to find happy foundations for the forms of behaviour that we prize in a full man and a fulfilled society? We have seen that human behaviour is characterised by a high internal delay in preparation for deferred action. The biological groundwork for this inaction stretches through the long childhood and slow maturation of man. But deferment of action in man goes far beyond that. Our actions as adults, as decision makers, as human beings, are mediated by values, which I interpret as general strategies in which we balance opposing impulses. It is not true that we run our lives by any computer scheme of problem solving. The problems of life are insoluble in this sense. Instead, we shape our conduct by finding principles to guide it. We devise ethical strategies or systems of values to ensure that what is attractive in the short term is weighed in the balance of the ultimate, long-term satisfactions.

And we are really here on a wonderful threshold of knowledge. The ascent of man is always teetering in the balance. There is always a sense of uncertainty, whether when man lifts his foot for the next step it is really going to come down pointing ahead. And what is ahead for us? At last the bringing together of all that we have learned, in physics and in biology, towards an understanding of where we have come: what man is.

Knowledge is not a loose-leaf notebook of facts. Above all, it is a responsibility for the integrity of what we are, primarily of what we are as ethical creatures. You cannot possibly maintain that informed integrity if you let other people run the world for you while you yourself continue to live out of a ragbag of morals that come from past beliefs. That is really crucial today. You can see it is pointless to advise people to learn differential equations, or to do a course in electronics or in computer programming. And yet, fifty years from now, if an understanding of man's origins, his evolution, his history, his progress is not the commonplace of the schoolbooks, we shall not exist. The commonplace of the schoolbooks of tomorrow is the adventure of today, and that is what we are engaged in.

And I am infinitely saddened to find myself suddenly surrounded in the west by a sense of terrible loss of nerve, a retreat from knowledge into – into what? Into Zen Buddhism; into falsely profound questions about, Are we not really just animals at bottom; into extra-sensory perception and mystery. They do not lie along the line of what we are now able to know if we devote ourselves to it: an understanding of man himself. We are nature's unique experiment to make the rational intelligence prove itself sounder than the reflex. Knowledge is our destiny. Self-knowledge, at last bringing together the experience of the arts and the explanations of science, waits ahead of us.

It sounds very pessimistic to talk about western civilisation with a sense of retreat. I have been so optimistic about the ascent of man; am I going to give up at this moment? Of course not. The ascent of man will go on. But do not assume that it will go on carried by western civilisation as we know it. We are being weighed in the balance at this moment. If we give up, the next step will be taken – but not by us. We have not been given any guarantee that Assyria and Egypt and Rome were not given. We are waiting to be somebody's past too, and not necessarily that of our future.

We are a scientific civilisation: that means, a civilisation in which knowledge and its integrity are crucial. Science is only a Latin word for knowledge. If we do not take the next step in the ascent of man, it will be taken by people elsewhere, in Africa, in China. Should I feel that to be sad? No, not in itself. Humanity has a right to change its colour. And yet, wedded as I am to the civilisation that nurtured me, I should feel it to be infinitely sad. I, whom England made, whom it taught its language and its tolerance and excitement in intellectual pursuits, I should feel it a grave sense of loss (as you would) if a hundred years from now Shakespeare and Newton are historical fossils in the ascent of man, in the way that Homer and Euclid are.

I began this series in the valley of the Omo in East Africa, and I have come back there because something that happened then has remained in my mind ever since. On the morning of the day that we were to take the first sentences of the first programme, a light plane took off from our airstrip with the cameraman and the sound recordist on board, and it crashed within seconds of taking off. By some miracle the pilot and the two men crawled out unhurt.

But naturally the ominous event made a deep impression on me. Here was I preparing to unfold the pageant of the past, and the present quietly put its hand through the printed page of history and said, 'It is here. It is now.' History is not events, but people. And it is not just people remembering, it is people acting and living their past in the present. History is the pilot's instant act of decision, which crystallises all the knowledge, all the science, all that has been learned since man began.

We sat about in the camp for two days waiting for another plane. And I said to the cameraman, kindly, though perhaps not tactfully, that he might prefer to have someone else take the shots that had to be filmed from the air. He said, 'I've thought of that. I'm going to be afraid when I go up tomorrow, but I'm going to do the filming. It's what I have to do.'

We are all afraid – for our confidence, for the future, for the world. That is the nature of the human imagination. Yet every man, every civilisation, has gone forward because of its engagement with what it has set itself to do. The personal commitment of a man to his skill, the intellectual commitment and the emotional commitment working together as one, has made the Ascent of Man.

Bibliography

CHAPTER ONE

Campbell, Bernard G., *Human Evolution: An Introduction to Man's Adaptations*, Aldine Publishing Company, Chicago, 1966, and Heinemann Educational, London, 1967; and 'Conceptual Progress in Physical Anthropology: Fossil Man', *Annual Review of Anthropology*, I, pp. 27–54, 1972.

Clark, Wilfrid Edward Le Gros, *The Antecedents of Man*, Edinburgh University Press, 1959.

Howells, William, editor, *Ideas on Human Evolution: Selected Essays, 1949–1961*, Harvard University Press, 1962.

Leakey, Louis S. B., *Olduvai Gorge, 1951–61*, 3 vols, Cambridge University Press, 1965–71.

Leakey, Richard E. R, 'Evidence for an Advanced Plio-Pleistocene Hominid from East Rudolf, Kenya', *Nature*, 242, pp. 447–50, 13 April 1973.

Lee, Richard B., and Irven De Vore, editors, *Man the Hunter*, Aldine Publishing Company, Chicago, 1968.

CHAPTER TWO

Kenyon, Kathleen M., *Digging up Jericho*, Ernest Benn, London, and Frederick A. Praeger, New York, 1957.

Kimber, Gordon, and R. S. Athwal, 'A Reassessment of the Course of Evolution of Wheat', *Proceedings of the National Academy of Sciences*, 69, no. 4, pp. 912–15, April 1972.

Piggott, Stuart, *Ancient Europe: From the Beginnings of Agriculture to Classical Antiquity*, Edinburgh University Press and Aldine Publishing Company, Chicago, 1965.

Scott, J. P, 'Evolution and Domestication of the Dog', pp. 243–75 in *Evolutionary Biology*, 2, edited by Theodosius Dobzhansky, Max K. Hecht, and William C. Steere, Appleton-Century-Crofts, New York, 1968.

Young, J. Z., *An Introduction to the Study of Man*, Oxford University Press, 1971.

CHAPTER THREE

Gimpel, Jean, *Les Bâtisseurs de Cathédrales*, Editions du Seuil, Paris, 1958.

Hemming, John, *The Conquest of the Incas*, Macmillan, London, 1970.

Lorenz, Konrad, *On Aggression*, Methuen, London, 1966.

Mourant, Arthur Ernest, Ada C. Kopeć and Kazimiera Domaniewska-Sobczak, *The ABO Blood Groups; comprehensive tables and maps of world distribution*, Blackwell Scientific Publications, Oxford, 1958.

Robertson, Donald S., *Handbook of Greek and Roman Architecture*, Cambridge University Press, 2nd ed., 1943.

Willey, Gordon R., *An Introduction to American Archaeology*, Vol. I, *North and Middle America*, Prentice-Hall, New Jersey, 1966.

CHAPTER FOUR

Dalton, John, *A New System of Chemical Philosophy*, 2 vols, R. Bickerstaff and G. Wilson, London, 1808–27.

Debus, Allen G., 'Alchemy', *Dictionary of the History of Ideas*, Charles Scribner, NewYork, 1973.

Needham, Joseph, *Science and Civilization in China*, 1–4, Cambridge University Press, 1954–71.

Pagel, Walter, *Paracelsus. An introduction to Philosophical Medicine in the Era of the Renaissance*, S. Karger, Basel and New York, 1958.

Smith, Cyril Stanley, *A History of Metallography*, University of Chicago Press, 1960.

CHAPTER FIVE

Heath, Thomas L., *A Manual of Greek Mathematics,* 7 vols, Clarendon Press, Oxford, 1931; Dover Publications, 1967.

Mieli, Aldo, *La Science Arabe*, E. J. Brill, Leiden, 1966.

Neugebauer, Otto Eduard, *The Exact Sciences in Antiquity*, Brown University Press, 2nd ed., 1957 ; Dover Publications, 1969.

Weyl, Hermann, *Symmetry*, Princeton University Press, 1952.

White, John, *The Birth and Rebirth of Pictorial Space*, Faber, 1967.

CHAPTER SIX

Drake, Stillman, *Galileo Studies*, University of Michigan Press, 1970.

Gebler, Karl von, *Galileo Galilei und die Römische Curie*, Verlag der J. G. Gotta'schen Buchhandlung, Stuttgart, 1876.

Kuhn, Thomas S., *The Copernican Revolution*, Harvard University Press, 1957.

Thompson, John Eric Sidney, *Maya History and Religion*, University of Oklahoma Press, 1970.

CHAPTER SEVEN

Einstein, Albert, 'Autobiographical Notes' in *Albert Einstein: Philosopher-Scientist*, edited by Paul Arthur Schilpp, Cambridge University Press, 2nd ed., 1952.

Hoffman, Banesh, and Helen Dukas, *Albert Einstein*, Viking Press, 1972.
Leibniz, Gottfried Wilhelm, *Nova Methodus pro Maximis et Minimis*, Leipzig, 1684.
Newton, Isaac, *Isaac Newton's Philosophiae Naturalis Principia Mathematica*, London, 1687, edited by Alexandre Koyré and I. Bernard Cohen, 2 vols, Cambridge University Press, 3rd ed., 1972.

CHAPTER EIGHT

Ashton, T. S., *The Industrial Revolution 1760–1830*, Oxford University Press, 1948.
Crowther, J. G., *British Scientists of the 19th Century*, 2 vols, Pelican, 1940–1.
Hobsbawm, E. J., *The Age of Revolution: Europe 1789–1848*, Weidenfeld and Nicolson, 1962; New American Library, 1965.
Schofield, Robert E., *The Lunar Society of Birmingham*, Oxford University Press, 1963.
Smiles, Samuel, *Lives of the Engineers*, 1–3, John Murray, 1861; reprint, David and Charles, 1968.

CHAPTER NINE

Darwin, Francis, *The Life and Letters of Charles Darwin*, John Murray, 1887.
Dubos, René Jules, *Louis Pasteur*, Gollancz, 1951.
Malthus, Thomas Robert, *An Essay on the Principle of Population, as it affects the Future Improvement of Society*, J. Johnson, London, 1798.
Sanchez, Robert, James Ferris and Leslie E. Orgel, 'Conditions for purine synthesis: Did prebiotic synthesis occur at low temperatures?', *Science*, 153, pp. 72–3, July 1966.

Wallace, Alfred Russel, *Travels on the Amazon and Rio Negro, With an Account of the Native Tribes, and Observations on the Climate, Geology, and Natural History of the Amazon Valley*, Ward, Lock, 1853.

CHAPTER TEN

Broda, Engelbert, *Ludwig Boltzmann*, Franz Deuticke, Vienna, 1955.

Bronowski, J., 'New Concepts in the Evolution of Complexity', *Synthese*, 21, no. 2, pp. 228–46, June 1970.

Burbidge, E. Margaret, Geoffrey R. Burbidge, Williarn A. Fowler, and Fred Hoyle, 'Synthesis of the Elements in Stars', *Reviews of Modern Physics*, 29, no. 4, pp. 547–650, October 1957.

Segrè, Emilio, *Enrico Fermi: Physicist*, University of Chicago Press, 1970.

Spronsen, J. W. van, *The Periodic System of Chemical Elements: A History of the First Hundred Years*, Elsevier, Amsterdam, 1969.

CHAPTER ELEVEN

Blumenbach, Johann Friedrich, *De generis humani varietate nativa*, A. Vandenhoeck, Göttingen, 1775.

Gillispie, Charles C., *The Edge of Objectivity: An Essay in the History of Scientific Ideas*, Princeton University Press, 1960.

Heisenberg, Werner, 'Über den anschaulichen Inhalt der quantentheoretischen Kinematik und Mechanik', *Zeitschrift für Physik*, 43, p. 172, 1927.

Szilard, Leo, 'Reminiscences', edited by Gertrud Weiss Szilard and Kathleen R. Winsor in *Perspectives in American History*, II, 1968.

CHAPTER TWELVE

Briggs, Robert W. and Thomas J. King, 'Transplantation of Living Nuclei from Blastula Cells into Enucleated Frogs' Eggs',

Proceedings of the National Academy of Sciences, 38, pp. 455–63, 1952.
Fisher, Ronald A., *The Genetical Theory of Natural Selection*, Clarendon Press, Oxford, 1930.
Olby, Robert C., *The Origins of Mendelism*, Constable, 1966.
Schrödinger, Erwin, *What is Life?*, Cambridge University Press, 1944; new ed., 1967.
Watson, James D., *The Double Helix*, Atheneum, and Weidenfeld and Nicolson, 1968.

CHAPTER THIRTEEN

Braithwaite, R. B., *Theory of Games as a tool for the Moral Philosopher*, Cambridge University Press, 1955.
Bronowski, J., 'Human and Animal Languages', pp. 374–95, in *To Honor Roman Jakobson*, I. Mouton & Co., The Hague, 1967.
Eccles, John C., editor, *Brain and the Unity of Conscious Experience*, Springer-Verlag, 1965.
Gregory, Richard, *The Intelligent Eye*, Weidenfeld and Nicolson, 1970.
Neumann, John von, and Oskar Morgenstern, *Theory of Games and Economic Behavior*, Princeton University Press, 1943.
Wooldridge, Dean E., *The Machinery of the Brain*, McGraw-Hill, 1963.

Index

Picture Credits

BBC Books would like to thank the following for providing photographs and illustrations, and for permission to reproduce copyright material. While every effort has been made to trace and acknowledge all copyright holders, we would like to apologise should there be any errors or omissions.

Page 24: Modern and fossil Nyala horns, Musée de L'Homme, Paris (Yves Coppens);
Page 27, left: The Taung child's skull, University of Witwatersrand, Johannesburg (Alan R. Hughes, permission of Prof. P. V. Tobias);
Page 27, right: Finger and thumb bones of *Australopithecus*, Mary Waldron;
Page 30: Computer graphic display of stages in evolution of the head, BBC;
Page 40: Rock painting, Erwin O. Christensen, by courtesy of Bonanza Books;
Page 43: Recumbent bison, Altamira, Michael Holford;
Page 54, top: Jericho skull, Ashmolean Museum, University of Oxford;
Page 54, bottom: The tower at Jericho tel, Dave Brinicombe;
Page 60, top left: Carpenter, National Museum, Copenhagen;
Page 60, top right: Clay treaty nail, The Trustees of the British Museum;

Page 60, bottom: Baker's oven, The Trustees of the British Museum;
Page 63: Carpenters at work with a bow-lathe, The British Library, Asia, Pacific and Africa Collections;
Page 67: Greek vase painting, The Trustees of the British Museum;
Page 79: Inca masonry at Machu Picchu, H. Ubbelohde Doering;
Page 84: The Great Mosque, Cordoba, Conway Library, The Courtauld Institute of Art, London. Photograph by A. F. Kersting;
Page 89: Masons at work, 13th century, from the Book of St Alban's, The Board of Trinity College Dublin;
Page 106, top left: Mask of an Achaean king, The Art Archive / Alamy
Page 106, top right: Gold dinar of Khusrau II, The Trustees of the British Museum;
Page 106, bottom left: Mochica puma, Museos 'Oro del Perú', 'Armas del Mundo', Fundación Miguel Mujica Gallo;
Page 106, middle right: Cast gold badge, Roynon Raikes / Victoria and Albert Museum, London;
Page 106, bottom right: Central input receiver in a pocket calculator, Paul Brierly;
Page 110, top: The furnace of the body, by Paracelsus, Wellcome Library, London;
Page 110, bottom: Earth, air and fire, by Paracelsus, S. Karger;
Page 117-118: Illustrations of atoms, BBC;
Page 121: Blind harpist, SSPL via Getty Images;
Page 126, left: Chinese block print of Pythagoras' theorem, The British Library Board (shelfmark Or.15378);
Page 126, right: Arabic version of Pythagoras' theorem, The British Library Board (shelfmark add.23387, f.28);

Pages 133 and 135: Illustration of architectural symmetry, BBC;
Page 141: Uccello's perspective analysis of a chalice, Gabinetto dei Disegni e Stampe, Uffizi, Florence, Italy / Alinari;
Page 149: Pages from *De Revolutionibus Orbium Coelestium*, World History Archive / Alamy;
Page 179: Halley's letter to Newton of 29 June 1686, by kind permission of the Provost and Scholars of King's College Cambridge;
Page 186: Computer graphic of the inversion of a sphere, BBC;
Page 205: A lightning conductor, from the Historical and Interpretive Collections of The Franklin Institute, Inc., Philadelphia, PA;
Page 207: A Wilkinson token, The Trustees of the British Museum;
Page 215: An elevator platform, BBC;
Page 220: Charles Darwin, akg-images;
Page 238: Dmitri Mendeleev in his last years, Ria Novosti / Science Photo Library;
Page 245: Dmitri Mendeleev in his last years, Novosti Press Agency;
Page 247: Illustration of Mendeleev's Patience, BBC;
Page 248: An early draft of Mendeleev's Periodic Table, Interfoto / Alamy;
Page 261: Exponential graphite-uranium pile, photo courtesy of Argonne National Laboratory;
Page 270: Röntgen's original X-ray plate, Deutsches Museum, Munich;
Page 271: X-ray diffraction pattern of a crystal of DNA, King's College London;
Page 273: Illustration of the Gaussian curve, BBC;
Page 280: Enrico Fermi, photo courtesy of Argonne National Laboratory;
Pages 282 and 283: The scientists' letter to President Roosevelt,

Argonne National Laboratory, by courtesy of Franklin D. Roosevelt Library;

Page 289: Gregor Mendel in 1865, Science Photo Library;

Page 297: Large chromosomes of onion skin cells, Brian Bracegirdle;

Page 305: Andrea Pisano, 'The Creation of Eve', akg-images / Orsi Battaglini;

Page 306: Cells of spirogyra, Arthur M. Siegelman;

Page 312: Leonardo, 'Child in Womb', Dennis Hallinan / Alamy;

Page 313: The author at home with Taung cast, D. K. Miller, Salk Institute;

Page 315: Dürer, 'Self-Portrait', Lehman Collection, New York

Page 323: The old city of Jerusalem, Jon Arnold Images Ltd / Alamy